ENERGY
AND THE
FUTURE

ALLEN L. HAMMOND
WILLIAM D. METZ
THOMAS H. MAUGH II

38132

AMERICAN ASSOCIATION FOR THE ADVANCEMENT OF SCIENCE
Washington, D.C.
1973

FOREWORD

Energy is the basic natural resource. To the consumer, it is the commodity he buys as gasoline, natural gas, and electricity. To engineers, it is the heat for industrial furnaces or the motive force that powers machinery. To the economist, it is a key ingredient in national prosperity. Without energy, man would be at the mercy of his environment, his cities uninhabitable, and many of the material goods he now uses unavailable.

Today the United States, with 6 percent of the world's population, consumes a third of the world's energy. The U.S. energy industry does business worth $100 billion a year, one tenth of the gross national product, and until recently could provide energy to meet every need. But all is not well. Industry spokesmen warn of a fuels crisis and point to declining domestic production of natural gas and oil, rising imports of these fuels, a sick coal industry, and a nuclear power industry beset by costly delays. Environmentalists oppose needless expansion of energy production and cite the pollution from power plants, offshore oil wells, and automobiles as prime threats to the quality of life. They single out a spendthrift attitude toward energy as a main cause of both energy shortages and pollution, and they warn that the environmental crisis will become more acute if energy consumption continues to rise unchecked, doubling and then quadrupling before the end of the century. Still other observers believe that environmental goals and energy requirements need not be incompatible. The real problem, in their view, is the management crisis in the federal government. Authority over energy matters is spread among 67 agencies and a dozen Congressional committees—leading to confusion evidenced by contradictory regulatory policies, lack of long-term planning, and an unbalanced research effort.

However one chooses to view the energy crisis, the stakes are high. It may seem inconceivable that this country's standard of living should decline, its industrial base move elsewhere, and its currency suffer continuing devaluation because of a lack of energy. Yet an overdependence on foreign energy supplies has been partly responsible for a similar chain of events in Great Britain. Energy played a crucial role in the decline of earlier cultures, too. During the Roman Empire, for example, the Mediterranean forests were cut down for fuel and lack of wood eventually forced smelters to move north to the German forests. By the end of the Empire, the Romans were forced to import iron from what had once been their provinces. The analogy with the United States in either case is more suggestive than exact; the present situation is both more complicated and more hopeful. But it may be significant that no new oil refineries have been built in this country for several years and that imports now account for more than a quarter of the U.S. oil supply.

Oil is a particularly crucial item in the energy budget. It provides all transportation fuels, heats many homes, and is replacing coal in many power plants. In all, oil accounts for nearly half of the U.S. energy supply. But domestic oil fields are beginning to run dry. The 10 million barrels of oil that they produce every day are not enough to keep pace with the growing demand. Despite removal of all limitations on production from the once vast Texas and Oklahoma fields and increased drilling in the Gulf of Mexico, U.S. production peaked two years ago. What oil remains in the ground is more expensive to find and to extract than oil from the huge Middle East reservoirs, which hold 75 percent of the world's known reserves. Yet to become dependent on imported oil may be to risk economic havoc. Estimates of the price tag for foreign oil run as high as $20 billion a year by the early 1980's. Even if the country could sustain its balance of payments in the face of such a drain on its capital, there is no guarantee that Middle Eastern countries will not use their oil wealth as an instrument of foreign policy, disrupting world money markets or exerting leverage on U.S. industries. Some observers fear that these countries might practice a direct form of energy blackmail in the event of a political crisis in the Middle East by shutting off the flow of oil and thus threatening this country's national security. These fears may be exaggerated, but massive oil imports are clearly something to be avoided if possible.

The truth of the matter is that we now have very few alternatives to oil. Coal is the most abundant fuel, but it is hard to mine, and, in many areas of the country, illegal to burn because of its sulfur content. Reserves of coal are huge, enough to last for hundreds of years. What is lacking is the technology to remove pollutants from coal and to convert it to clean fuels. Coal mining as it is now practiced is so hazardous that a labor shortage exists because few men can be induced to accept the risks. Automated machinery that would make deep mining safer does not yet exist. Neither do strict laws to insure the reclamation of strip-mined areas.

Natural gas is an ideal fuel, but one that is in short supply. Use of this priority fuel has been overpromoted; most of natural gas is burned under the boilers of electric companies and used for other industrial needs that could be met by coal or oil. Now the supply of natural gas is nearly exhausted. Reserves are rapidly declining, and the price of gas is so low that oil companies no longer drill for it when they can drill for oil instead. If consumption of natural gas continues at the present rate and exploration does not pick up, the United States may burn its last molecule of domestic natural gas within 20 years.

The choices have narrowed to the point where the only new source of energy available now is nuclear power. But despite their rapid proliferation, nuclear generating plants provide a meager 1 percent of this country's energy. Use of nuclear power cannot expand quickly enough to fill the gap left by unavailability of traditional fuels. For the immediate future—at least the next few years—there is no escape from the necessity of importing ever larger quantities of oil or of cutting energy consumption drastically.

How did it happen that this country finds itself seemingly unprepared for the future? How did the energy crisis catch us unaware? One answer is that the system worked so well until the mid-1960's that no one paid any attention to where things were going. In both public and private decision making, most people acted as if ever increasing amounts of energy at ever lower prices would always be available —and the fallacy of this assumption is now apparent.

Existing policies are also part of the problem; by restricting imports of foreign oil, this country faces shortages of heating oil and gasoline at a time when there is a surplus of crude oil on the world market. The situation is no more rational in regard to conserving energy. Half a dozen federal agencies, from the Tennessee Valley Authority to the Rural Electrification Program, are specifically chartered to promote energy consumption. Key regulatory opinions from the Federal Power Commission and the Interstate Commerce Commission favor increased consumption rather than conservation. The result is that the demand for energy has continued to grow rapidly with neither conservation policies to slow it nor enough imported oil to prevent shortages.

A more cynical view lays the fault for this country's present energy difficulties to the economic greed of the oil industry and its political allies. This view may be overstated, but it is clear that the economic self-interest of the oil companies does not always coincide with that of the consumer or that of the nation as a whole. Oil import quotas, for example, have had the effect of guaranteeing the oil industry a minimum price for its product and of hastening the day when domestic reserves will be gone.

Last winter's shortages of heating oil provide a case study of industry's part in creating the energy crisis. Under Phase 2 price ceilings, refineries found it more profitable to make gasoline than heating oil and delayed converting their production far past their usual date. Despite sagging inventories of heating oil, major oil companies assured the White House Office of Emergency Preparedness in September, 1972, that enough oil to provide for the winter would be available. Then when shortages developed, in spite of a mild winter in much of the country, the industry petitioned state regulatory authorities to relax environmental standards and permit the burning of oils with higher sulfur content. At the same time, the American Petroleum Institute launched a major public relations campaign in which it spent $3 million to convince the country that the oil industry was doing everything it could to avert a crisis. What the ads did not point out was that some of the major oil companies were contributing to the shortage by hoarding heating oil overseas and thus drying up the world market in this commodity. Small independent oil importers, who hold a share of the available import permits, were consequently unable to buy heating oil abroad. As a result, less heating oil was brought into the country in January 1973 than in the same period a year earlier, despite higher import quotas for 1973.

The federal government has also lagged in taking steps to insure a supply of oil. A cabinet task force in 1969 urged that oil import quotas be eliminated and a system of tariffs be instituted in their place. Import restrictions, the task force found, were costing American consumers $6 to $7 billion a year by maintaining domestic oil prices at artificially high levels. Since 1969, demand for oil has risen sharply as electric companies in many major cities have converted from coal to residual oil to meet increasingly strict air pollution standards. Ad hoc adjustments of the import quotas year by year were not enough to keep up with changing patterns of fuel use, but did have the effect of discouraging long-term commitments by industry for supplies and new refineries.

Doing away with import quotas will help to ease the present artificial shortages of oil and oil products. Indeed, last winter's oil shortages, as former White House energy advisor S. David Freeman put it, "could have been averted with the stroke of a pen." Far more genuine problems are the shortage of natural gas in many parts of the country, the inability to burn coal without relaxing environmental standards, and the lack of other energy options. Equally serious are policy questions about the price of natural gas and the priorities for its use, about the environmental consequences of permitting offshore drilling for oil and gas in coastal waters, and about the role of the federal government in energy matters. What seems to be needed, at the least, is a massive program of research and development to broaden the choices for energy and a firm national commitment to energy conservation.

That this country cannot afford to be without more energy options is the general thesis of this book. But policy questions such as those mentioned above do not receive much attention here. Instead, this book seeks to discover and to assess the technologies and research developments that will be the basis for future energy policies. My co-authors and I hope that it will prove a useful guide to energy research.

ALLEN L. HAMMOND

March 12, 1973

CONTENTS

V. ENERGY CONSERVATION

VI. ENERGY POLICY

LIST OF ILLUSTRATIONS

I.
ENERGY FROM FOSSIL FUELS

"There are two things wrong with coal today. We can't mine it and we can't burn it."

—S. DAVID FREEMAN,
Director, Ford Foundation Energy Policy Project

1.

ENERGY FROM FOSSIL FUELS

Coal is unquestionably the fossil fuel of the future.

For the present, however, the U.S. energy industry is firmly structured around petroleum, and it is likely to remain that way for some time. Oil and natural gas together now account for some 77 percent of U.S. energy consumption—a fact that contrasts sharply with estimates that 89 percent of all fossil fuel reserves are in the form of coal. But the petroleum industry has grown with the U.S. economy, and explanations of its current dominance are not hard to find.

In the first place, fluids are easier to handle and use than are solids. This is particularly true for mobile combustion sources, which account for more than a quarter of all energy use. The massive post-war growth of the high-pressure gas pipeline network, moreover, made natural gas readily available in all sections of the country. And more recently, air pollution regulations in many areas have forced the substitution of natural gas and low-sulfur residual fuel oil for high-sulfur coal.

Oil, in particular, is also highly versatile. Technology is available for separating or reforming it into a wide variety of liquid fuels, for removing its sulfur, and for converting some of its components to synthetic natural gas or even to hydrogen. Petroleum production is large and growing, and massive amounts of money have been invested in refineries and other processing plants; sheer inertia alone would enable petroleum to maintain its dominance of the energy market for as long as it is available.

Adequate supplies will be available for many years, particularly from the great reserves in the Middle East, but these supplies will be obtained only at increasing prices (although Middle Eastern crude is still cheaper than domestic oil), with lessened national security, and with great damage to the U.S. balance of payments. Eventually, perhaps between 1985 and 2000, these stresses will combine with decreasing reserves and increasing demand from other sectors of the world to force a shift to fuels that are more abundant or that can be produced domestically.

One result of this shift could be production of oil from shale. Oil shale is a laminated marlstone rock containing a tarlike organic material called kerogen. When heated to 450 to 600°C, kerogen releases vapors that can be converted to raw shale oil, a black, viscous substance that can, in turn, be refined into petroleum products. U.S. oil shale reserves may be as much as eight times as large as petroleum reserves.

Technology for shale oil production has been thoroughly tested in the laboratory and is, in fact, simpler than that for coal gasification or liquefaction. But further development has been hindered by three major problems: cost, environmental disruption, and uncertainties about ownership of the shale.

Price estimates vary, but it seems likely that shale oil would cost a minimum of $4.50 per barrel to produce, or roughly $1.00 a barrel more than domestic crude oil. A major fraction of this price is processing costs resulting from the low concentration of kerogen in oil shale. Even in the best deposits, more than one and a half tons of rock must be processed for each barrel of oil produced. Disposal of the processed rock, furthermore, could create major environmental problems, since a large commercial refinery would produce as much as 1.5 million tons per day. Damage to the environment could thus be much greater than in strip mining of coal, for example.

Another difficulty is that more than 80 percent of U.S. oil shale is located on federal lands. The few companies that have shown interest in shale oil have not found the government's terms for leasing this land sufficiently attractive to warrant a large investment. But even highly favorable leasing terms would probably not be sufficient to overcome the first two problems, and it is unlikely that shale oil development will begin for at least another 10 years. The projected $4.50 cost could, however, serve as a price ceiling for imported oil.

For all practical purposes, then, a shift away from petroleum will mean a shift toward coal, which is widely available and has a long history of use. But that shift will be neither easy nor inexpensive, for there are many problems associated with coal use. The high sulfur content of much U.S. coal and environmental damage from mining have been the most widely discussed of these problems, but a variety of others—particularly the composition of the coal itself—have proved to be equally troublesome.

There are four major grades of U.S. coal: anthracite, bituminous, subbituminous, and lignite. These classifications reflect the amounts of fixed carbon (involatile) organic compounds, volatile organic compounds, moisture, and ash in each grade. In general, higher fixed carbon and lower moisture contents indicate a greater energy content and a higher quality of coal (Table 1).

Table 1. Proximate analyses of typical samples from the major classes of U.S. coal.

Type	Origin	Proximate analysis (%)				Heating value (Btu per pound)
		Moisture	Volatile matter	Fixed carbon	Ash	
Anthracite	Pa.	4.4	4.8	81.8	9.0	13,130
Bituminous						
Low-volatile	Md.	2.3	19.6	65.8	12.3	13,220
Medium-volatile	Ala.	3.1	23.4	63.6	9.9	13,530
High-volatile	Ohio	5.9	43.8	46.5	3.8	13,150
Subbituminous	Wash.	13.9	34.2	41.0	10.9	10,330
	Colo.	25.8	31.1	38.4	4.7	8,580
Lignite	N. Dak.	36.8	27.8	30.2	5.2	6,960

[Source: Bureau of Mines]

Anthracite is the top grade; it generally contains 86 to 98 percent fixed carbon and less than 2 percent moisture. It is hard, easy to handle, contains very little sulfur, burns cleanly, and has an energy content as high as 16,000 Btu per pound. Its use has been declining, however, because it is the most expensive coal and it is in relatively short supply. The principal reserves of anthracite are in Pennsylvania, but nearly half those reserves have already been mined. The remainder constitutes less than 2 percent of total coal reserves in the U.S.

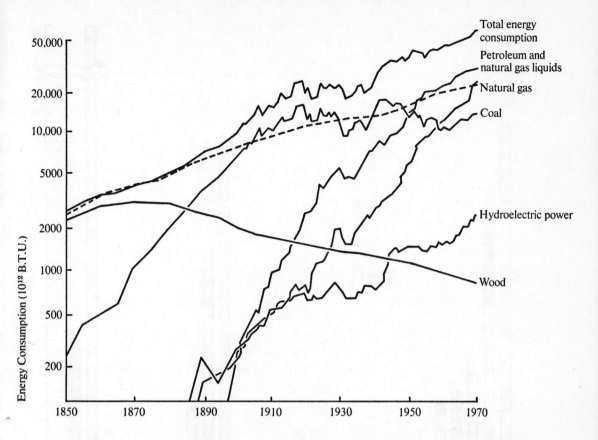

FIGURE 1. U.S. energy consumption (in trillions of British thermal units) since 1850. [*Source: From "An Agenda for Energy," by Hoyt J. Hottel and J. B. Howard. Copyright © Technology Review.*]

Bituminous, the most widely used, is a softer coal that contains between 50 and 86 percent fixed carbon. It is relatively easy to handle, burns with only moderate production of smoke, and has an energy content of 11,000 to 15,000 Btu per lb. It also has a very high sulfur content. More than 70 percent of U.S. bituminous reserves contain at least 1 percent of sulfur and 43 percent contain more than 3 percent sulfur—making their use undesirable (or illegal) for most applications. Nonetheless, the bituminous coal fields of the East and North Central regions have in recent years accounted for more than 70 percent of U.S. coal production.

Subbituminous and lignite coals account for about 56 percent of U.S. coal reserves, but they are of relatively low quality. Subbituminous generally contains 40 to 60 percent fixed carbon, 20 to 30 percent moisture, and as much as 15 percent ash; it has an energy value of 8,000 to 12,000 Btu per lb. Lignite contains less than 40 percent fixed carbon and as much as 40 percent moisture, and has an energy value of 5,500 to 8,000 Btu per lb. Both types are relatively free of sulfur, but produce large quantities of fly ash on combustion.

The major reserves of these two grades lie in the Northern Plains and Rocky Mountain states and have been virtually untouched because there is no industry there to use them and it is too expensive to ship them. Lignite, and subbituminous to a lesser extent, dries out on exposure to air and crumbles; it must therefore be shipped under carefully controlled conditions, and is difficult to store. These

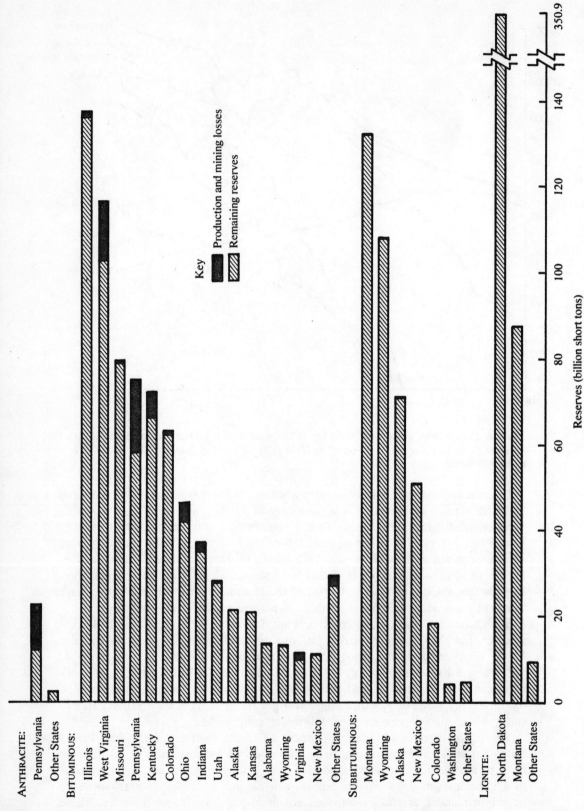

FIGURE 2. Estimated original and remaining coal reserves in the United States.
[*Source: Bureau of Mines Information Circular 8312.*]

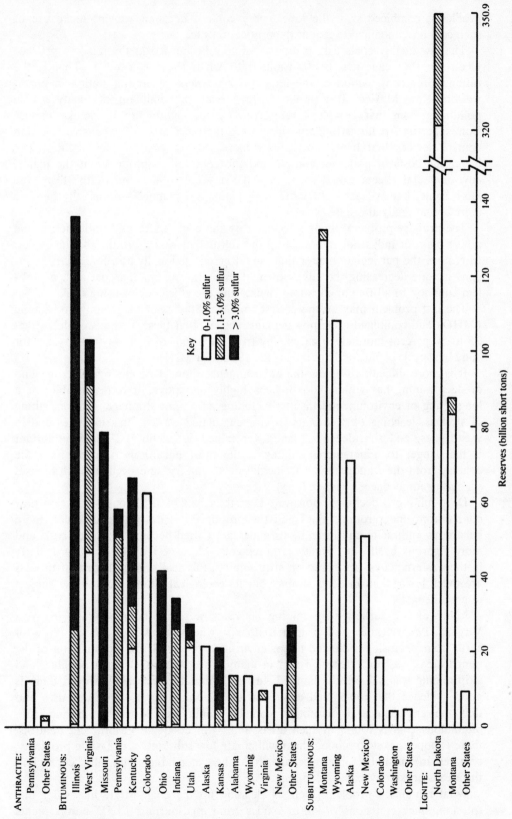

FIGURE 3. Sulfur content of estimated coal reserves in the United States.
[*Source: Bureau of Mines Information Circular 8312.*]

problems, combined with the low energy content per unit weight, make use of such coal uneconomical except at the production site.

The low quality coals can, of course, simply be burned at the mine to produce electricity that can then be transmitted to wherever it is needed: That is the rationale for construction of the large Four Corners generating station in northwestern New Mexico. But public outrage over the emission of highly visible pollutants from that complex—reinforced by the sighting of its smoke plumes from space by Apollo astronauts—has forced the installation of expensive pollution control devices that have substantially increased the cost of the facilities. The increasing costs of pollution control and the continuing uproar about the plants' environmental effects would seem to make a wider application of the technique improbable, but at least eight more such plants are planned—principally because there are no ready alternatives.

The primary problem, then, is to clean up the coal before its combustion at the power plant or industrial site—to trap the sulfur that would produce sulfur oxides, to remove the particulate matter that would be emitted as fly ash. Subsidiary problems include decreasing the environmental disruption of mining coal and, a problem common to all fuel applications, increasing the efficiency of using coal.

The last problem may be the easiest to solve. Studies of magnetohydrodynamic (MHD) and combined cycle power plants suggest that these methods could extract 25 to 50 percent more electrical energy from each unit of fuel—within 5 years for combined cycle plants, perhaps during the 1980's for MHD.

It is more difficult to assess the environmental impact of mining. Underground or deep mining has without doubt been highly disruptive. It is responsible for a long string of environmental insults, including acid mine drainage, erosion, silting of streams, leaching of pollutants from accumulated wastes, fires in mines and in waste piles, and subsidence of land over mined-out cavities. Equally important is the danger to miners; coal mining is the most hazardous occupation in the country from the standpoint of accidental death, injury, and occupational diseases such as pneumoconeosis (black lung).

Deep mining is declining, however. Less than half of U.S. coal production now comes from underground mines and the proportion is decreasing as older, more hazardous anthracite and bituminous mines are closed because of higher costs and more rigorous health and safety requirements. Most new operations, particularly in the western states, are based on strip mining; the health hazards of that process are much lower than in deep mining, but its ecological effects are still the subject of great debate.

Many of the earliest strip mining activities were in the mountains of West Virginia, Pennsylvania, and eastern Kentucky, where removal of the top layers of earth from steep slopes created major environmental problems. Reclamation of the mined areas was not attempted and in many cases was not deemed possible, and strip mining was universally damned. In less mountainous areas, however, there is good evidence that the mined areas can be reclaimed for a variety of uses and that the environmental disruption can be minimized. Most lignite and subbituminous coal is found in such areas and, with adequate government controls, there is hope that strip mining them will create few substantive or lasting problems; some environmentalists, however, are afraid the water table might fall if caprock in the area is broken by extensive mining.

What remains a problem, then, is cleaning up the coal. MHD may offer one solution, since emissions of sulfur oxides from coal-burning MHD power plants are expected to be low. Designing these plants to withstand corrosion from coal ash is quite difficult, but the possibility of providing an environmentally acceptable way to burn coal at an efficiency greater than that of conventional steam cycles makes MHD an attractive option.

The principal thrust of coal research, however, has been conversion of solid coal to another form, gas or liquid, from which undesirable components can be removed easily. This effort has been directed along three major routes: gasification, pyrolysis, and solvent refining. With the exception of one gasification process developed more than 30 years ago, none of these has shown sufficient success or received sufficient funding for commercialization.

Solvent refining—also called hydroprocessing or hydrorefining—involves the use of a coal-derived heavy aromatic solvent, generally under a moderate hydrogen pressure, to dissolve the organic components of coal. The solution is filtered to remove ash and insoluble organic materials and fractionated to recover the solvent. Small quantities of hydrocarbon gases and light liquids are produced, but the principal product is a heavy organic material called solvent-refined coal. This material has a melting point of about 180°C and contains less than 0.1 percent ash; its uniformly high heating value is about 16,000 Btu per pound regardless of what coal is used, and it can be used in place of coal with only minor modification of combustion equipment. The process removes all inorganic sulfur from the coal and 60 to 70 percent of the organic sulfur.

Solvent refining was developed for the Department of the Interior's Office of Coal Research (OCR) by two subsidiaries of Gulf Oil Corporation. Sufficient process technology for construction of a pilot plant has been available for at least 5 years, but no funds were available. OCR recently received the necessary funds for construction of a pilot plant in Tacoma, Washington, however, and it should begin operation within two years.

One interesting variation of solvent refining is a liquefaction process developed for OCR by Consolidation Coal Company, Pittsburgh, Pennsylvania. During the extraction period in this process, the organic material is sufficiently hydrogenated for it to remain in liquid form. Part of the material is recycled as fresh solvent; the remainder can be treated with a catalytic hydrocracking (H-Oil) process developed by Hydrocarbon Research Inc., New York City, to produce a synthetic crude oil.

The liquefaction technology, without the H-Oil process, was tested in a Cresap, West Virginia, pilot plant between 1967 and 1970. Mechanical problems prevented a satisfactory demonstration of the technique, but limited data showed that operation of the plant was consistent with laboratory results. Consolidation Coal has proposed revamping the plant to incorporate the H-Oil process and other improvements, but OCR has not received authorization or appropriations to fund the project.

Pyrolysis is a more highly developed process in which coal is heated in the absence of oxygen to produce a mixture of char, oil, and low-Btu gas. With minor variations, the process is the same as that described for pyrolysis of organic wastes in Chapter 5. FMC Corporation has operated a fluidized bed pyrolysis pilot plant (Project COED, for Char-Oil-Energy-Development) in Princeton, New Jersey, since 1970, and OCR has termed the plant and the project highly successful.

Although additional improvements are being incorporated in the COED plant, the projected costs for oil and char produced by pyrolysis are still too high to justify extension of the project to a larger demonstration plant. Only when the cost of pyrolytic oil is lowered or the cost of domestic crude oil rises dramatically will such a process become commercially viable, and neither prospect seems likely within this decade.

In fact, the only fossil fuel whose price is rising rapidly enough to justify use of a synthetic product is natural gas. Imported liquefied natural gas is expected to cost nearly $1.50 per million Btu (1000 standard cubic feet) in the very near future, compared to an average domestic price of less than 50 cents per million Btu. That price difference has kindled a strong interest in gasification, and work is progressing rapidly in two areas: production of a high-Btu synthetic natural gas that could be used to replace natural gas in existing pipelines, and production of a low-Btu gas—often called power gas—for use in conjunction with combined cycle and MHD power plants to generate electricity. Gasification thus stands by far the best chance of being the first commercial coal conversion technology.

2.

GASIFICATION
A Rediscovered Source of Clean Fuels

Natural gas is one of our most precious energy resources. Not only is it the cheapest and most versatile fossil fuel available but, perhaps most important, it is also the least polluting. As industry and utilities have been forced to meet ever more stringent air pollution regulations, demand for natural gas has surged. Domestic consumption rose to 22.7 trillion standard cubic feet (scf) in 1971, an increase of more than 73 percent since 1961.

This increased demand has pushed natural gas producers and distributors to the limits of their capacity. Many distributors no longer accept new customers, and some are even restricting the amounts of gas available to their present customers. Others are meeting their requirements only by importing liquefied natural gas at prices (about $1 per 1000 scf) that are generally more than twice the U.S. rate. Proven U.S. reserves, furthermore, are sufficient for only about 13 more years at the present rate of consumption (although the American Petroleum Institute projects that 66 percent of U.S. reserves has not been discovered yet). Some studies indicate that by 1980, demand could outstrip supply by 20 billion scf per day.

Frequently, however, crisis spawns new technological developments. The potentially massive natural gas shortage has thus brought to the forefront a host of both old and new technologies for producing synthetic natural gas (SNG) from other fossil fuels. Industrial plants incorporating this technology should begin contributing to gas supplies within 2 years, and by 1985 may contribute as much as 15 percent of the total.

The basic chemistry of gasification is simple. Carbon from coal or naphtha—the petroleum fraction with a boiling point between 125 and 240°C—is combined with water at high temperature to form methane, the principal constituent of natural gas. The overall reaction requires several steps, however, and is much more complex.

Naphtha gasification is considerably simpler than coal gasification and is in a much more advanced state of development. Three different naphtha processes have been commercialized: the catalytic rich gas (CRG) process developed by the British Gas Council, the methane rich gas (MRG) process developed by Japan Gas Company, and the Gasynthan process developed by West Germany's Lurgi-Gesellshaft für Mineralöltechnik GmbH and Badische Anilin-und Soda-Fabrik AG. These processes are very similar in concept; the main difference is in the catalysts used.

In all three processes, vaporized naphtha is superheated under pressure and catalytically desulfurized. The sulfur-free vapor is then reacted with superheated

Table 2. Cost of various clean fuels (in cents per million Btu).

Fuel	Average selling price
Natural gas—at wellhead*	
Texas	16.1¢
Louisiana	20.2
Alaska	23.8
U.S. average	18.2
Natural gas—at New York City gate	35 to 45
Liquid natural gas—East coast	80 to 100
Low-sulfur fuel oil—East coast	65 to 75
SNG from light petroleum fractions	110 to 120
SNG from coal—Lurgi	105 to 115
SNG from coal—U.S. processes**	70 to 95

*January 1972. **Assuming use of strip-mined western coals.
[Sources: Bureau of Mines, Stanford Research Institute]

steam at high temperature (500 to 540°C) and pressure (34 atm) to form "synthesis gas"—a mixture of methane, hydrogen, and carbon monoxide—and carbon dioxide. This gas is then subjected to a catalytic methanation in which three molecules of hydrogen are combined with one of carbon monoxide to form more methane. After carbon dioxide and water are removed, the product gas is about 95 to 98 percent methane with an energy content between 980 and 1035 Btu per scf—the same as that of natural gas.

Each of the three processes has been proved in small-scale plants abroad, but there are as yet no commercial plants in this country. American utilities and gas producers have, however, ordered at least 25 such plants with a total capacity of more than 3 billion scf per day, and the first of these should begin production in less than 2 years. At least 17 of these plants will be based on the CRG process.

But these plants may be no more than a stopgap to the natural gas crisis, for they also face a future feedstock shortage. Almost all of the naphtha to be used in these plants must be imported and is thus subject to both rigorous U.S. import quotas and the caprice of producing countries. Increasing quantities of naphtha are also being used as raw material by the world chemical industry, and this competition will further limit the amount available for gasification. Gas producers will thus be forced to seek more abundant supplies of raw material, and the sole suitable alternative is coal.

The United States has massive reserves of coal—enough to last more than 500 years at the current rate of consumption. Yet much of it contains such high quantities of sulfur that environmental regulations prevent its use, or the energy content is so low that transportation of the coal to power plants is economically prohibitive. Conversion of this coal to sulfur-free SNG may thus be the only feasible way to use it.

Gas made from coal was once widely used in the United States: it has been only about 20 years, in fact, since U.S. utilities shifted from coal gas to natural gas. The processes used to manufacture coal gas were crude and inefficient, however; gas output from plants using these processes was low, and the cost was high. The product usually contained more than 50 percent hydrogen and carbon monoxide, and the energy content was never higher than 450 Btu per scf. If more efficient production methods can be developed, low-Btu gas will very likely find widespread use for onsite generation of electrical power, but it is not interchangeable with natural gas, and it cannot be transported economically over long distances.

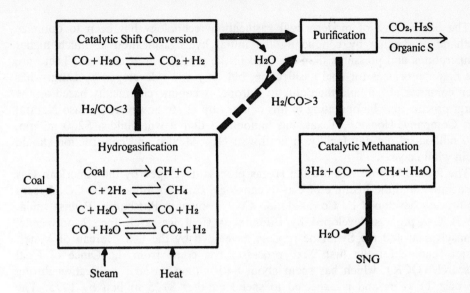

FIGURE 4. A generalized flow chart for the gasification of coal. The scheme is similar for naphtha gasification, except that desulfurization occurs first.

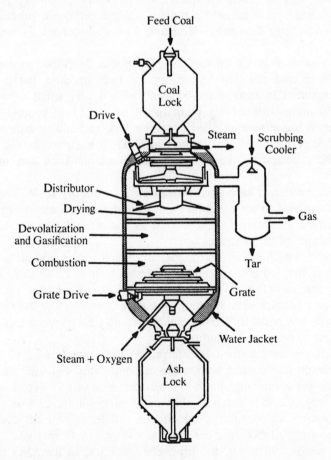

FIGURE 5. Cross-section of the Lurgi gasifier, the only coal gasifier now in commercial use. [*Source: Lurgi Mineralöltechnik GmbH.*]

The basic thrust of modern coal gasification technology has been to optimize methane production by conducting the initial hydrogasification at much higher temperatures and pressures than are used in the production of coal gas. There are five major processes for coal gasification, but only one—developed by Lurgi—has been commercialized, and then only in Europe. A commercial facility based on the Lurgi process may be operating in this country by 1976, however; El Paso Natural Gas Company, Houston, Texas, has announced that it will build a $250 million, 250 million-scf-per-day facility in northwest New Mexico and construction should begin within a year.

The four U.S. processes are the Hygas process developed by the American Gas Association and the Institute of Gas Technology, Chicago, Illinois; the CO_2 Acceptor process developed by Consolidation Coal Company, Pittsburgh, Pennsylvania; the Bi-Gas process developed by Bituminous Coal Research Inc., Monroeville, Pennsylvania; and the Synthane process developed by the U.S. Bureau of Mines. Major funding for the first three processes has come from the Office of Coal Research (OCR), which has spent about $40 million on coal gasification during the past 11 years and is expected to spend another $125 million by 1975. The Hygas and CO_2 Acceptor processes are being tested in pilot plants; construction has begun on a pilot plant for the Bi-Gas process; and the design work for a Synthane pilot plant is nearly complete. Several other promising processes are under development—most notably a molten sodium carbonate process conceived by the M. W. Kellogg Company, Houston, Texas—but none is near the pilot plant stage.

The Lurgi process is based on technology that was developed in the 1930's for the production of coal gas. The technology has been updated, but the reactor is still a low-pressure (28 atm), fixed-bed unit with a very small capacity. Consequently, a 250 million-scf-per-day plant based on the Lurgi process requires 31 gasifiers, whereas a comparable plant based on U.S. technology requires only two or three. Although the initial cost of the two plants is comparable, the Lurgi plant would thus be expected to require much more maintenance and may also be less reliable.

All four U.S. processes have been proved on a laboratory scale. The pilot plant programs are designed to solve the mechanical and chemical engineering problems inherent in scaling a laboratory reaction up to commercial size and to assess the economic feasibility of the processes.

The basic unit in each process is the gasifier, which operates at pressures ranging from 20 to more than 70 atm and temperatures as high as 1500°C. In the generalized reaction scheme (Fig. 4), coal is admitted to the reactor under pressure and brought into contact with synthesis gas at temperatures of 600 to 800°C to drive off volatile components. These components may be converted to methane or collected for use by the chemical industry.

The devolatilized coal is then transferred to the second stage of the reactor, where it is brought into contact with steam at temperatures greater than 900°C to form synthesis gas containing 40 to 65 percent methane. If necessary, the synthesis gas is subjected to a catalytic shift conversion similar to that employed in naphtha gasification. Carbon dioxide, hydrogen sulfide, organic sulfides, and water vapor are then stripped from the gas, and it is subjected to catalytic methanation. After further removal of water, the product is identical to the SNG produced by naphtha gasification.

The principal differences between the processes include the manner in which

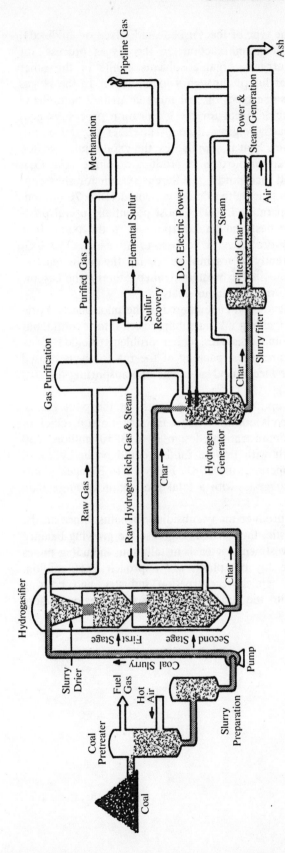

FIGURE 6. Flow chart for the Hygas process for coal gasification. Hygas is believed by many to be the most promising of the four major U.S. gasification processes. [*Source: Institute of Gas Technology.*]

the coal is admitted to the gasifier, the type of reactor bed used (fixed or fluidized), and the source of heat for the gasification reaction. In the Hygas process, for example, coal is admitted as a slurry in organic solvents, while in the other processes it is admitted as powder or lumps through a lock-hopper. In the Hygas and Synthane processes, moreover, some types of coal must be treated beforehand to prevent agglomeration in the reactor. Heat for the gasification process is generally provided by burning char—partially gasified coal—in either oxygen or air. In the CO_2 Acceptor process, however, heat is supplied by the exothermic reaction of calcined dolomite (CaO) with carbon dioxide to produce dolomite ($CaCO_3$).

Each pilot plant, furthermore, will incorporate a different system for shift conversion, cleanup, and methanation so that by the time a proposed 60 to 70 million-scf-per-day demonstration plant is operating in 1976, most potentially useful processes will have been tested, and the best can be incorporated in the plant. It is quite possible, though, that no one process will be best in all cases. The CO_2 Acceptor process, for example, currently seems best suited for the more reactive lignite and subbituminous coals found in the western United States, whereas the Bi-Gas process appears best for eastern bituminous coals.

Numerous engineering problems have been encountered in the pilot plants. Foremost among them is the problem of melding all the subprocesses into a continuous one operating at high temperatures and pressures. Other problems include structural failures and corrosion at high pressures, packing of the reactor beds caused by agglomeration of the coal, and the mechanical problems of transporting the coal through the system.

Despite these problems, OCR is optimistic about progress at the pilot plants. If current schedules and funding levels continue, says Edward Larson, chief of OCR's division of contracts and administration, design of a $250 million, 250 million-scf-per-day plant (to be built with private funds) could begin by 1977, with the first such plant going into operation by 1981. From 12 to 37 such plants could be in operation by 1985, he suggests, with a total production of more than 9 billion scf per day.

Construction of such plants could produce many subsidiary benefits. Since all the materials used in the plant are domestic, they would help ease the growing balance of payments deficit. Total capital investment for each installation, including mines to produce 15,000 tons of coal per day per plant, will approach $350 million, bringing a modest boom to the chemical construction industry and pouring unprecedented amounts of capital into the economically depressed regions where most coal resources are located.

3.

POWER GAS AND COMBINED CYCLES
Clean Power from Fossil Fuels

In the search for new ways to produce clean electric power from fossil fuels, generating systems that combine gas and steam turbines are playing a central role. Combined gas and steam cycle systems offer a cheap and—more important—an immediately available way to meet increasing demand for electricity in the next few years. Approximately 15 units with a total capacity of nearly 4500 megawatts have been ordered by U.S. companies in the past 18 months, and the first is due to be delivered in mid-1973. Combined cycle systems are attractive to the utility industry now because of delays in construction of nuclear facilities, difficulties in siting large generators, and pressures to use clean fuel. They may be even more attractive in the future because they promise to be more efficient than conventional stations. The new units are designed to burn natural gas and oil distillates, but could also burn other gasified fuels.

As natural gas and low-sulfur oil become scarce, combined-cycle systems may be the key to clean production of electricity from coal. Coal gasification to make power gas appears to be one of the cheapest ways to eliminate sulfur emissions from coal combustion, and the most efficient way to burn power gas would be a combined-cycle system.

No processes for making power gas from coal have been developed by U.S. companies. A clear distinction must be made, however, between power gas, which has a rather low heating value [about 150 Btu per standard cubic foot (scf)], and synthetic natural gas, which has a high heating value (about 1000 Btu per scf). Processes for making synthetic natural gas from coal have been funded by the Office of Coal Research (OCR) since 1961.

Unlike natural gas or its synthetic substitutes, power gas has such a low heating value that it cannot be economically transported very far; it must be used near the site of production. However, the synthesis of power gas is far simpler than the synthesis of natural gas, a process that one engineering consultant has called the toughest problem in chemical development he has ever known. The only power station using a combined-cycle system fueled with power gas (low-Btu gas from coal) has just begun operating at Lünen, Germany, so there is no indication yet just how expensive such a process will be.

Gas turbines that could be used for power generation were an offshoot of the development of turbines for military aircraft in World War II. The performance of aircraft turbines continues to improve as metallurgists find metals that withstand ever higher temperatures. New developments in the turbines of military aircraft become available to civilian aircraft 2 or 3 years later, and appear in the design of stationary turbines for industry about 5 years later. The turbine engines in the military aircraft of World War II could withstand maximum inlet temperatures of only about 500°C, but, by using new metals and new techniques for cooling the turbine blades, aircraft turbines can now operate at almost 1200°C (Fig. 7). Turbines for electrical power generation now operate at 1000°C.

The advantage of a combined-cycle system is that it has the potential of greater efficiency than either a gas or steam turbine alone. Much of the heat entering a gas turbine is wasted when the exhaust gases escape at relatively high temperatures (typically 445°C for a turbine with a 1000°C inlet temperature). These losses lower the efficiency for converting fuel energy into electricity. (Generators powered by gas turbines alone have efficiencies of about 25 percent and have been used for several years by U.S. utility companies to meet power requirements at times of peak loads.) If the hot exhaust gases are channeled into a boiler rather than allowed to escape, more electricity can be produced. The efficiency of combined gas and steam turbine systems is about 39 percent—comparable to the efficiency of the best existing steam-power installations.

For any gas turbine, the efficiency of power production increases with the temperature of the hot gases that turn it. Efficiency gained by operating a combined-cycle system at high temperatures is even greater. According to a report by the United Aircraft Research Laboratories, temperatures of 1220°C should be attainable by the mid-1970's, and 1440°C by the early 1980's. At the latter temperature, the efficiency of the combined cycles should be 50 percent.

Efficiency may be improved even further by adding another cycle. Systems that add an extra turbine at the high temperature are called topping cycles, and those that add an extra turbine at the low temperature are called bottoming cycles.

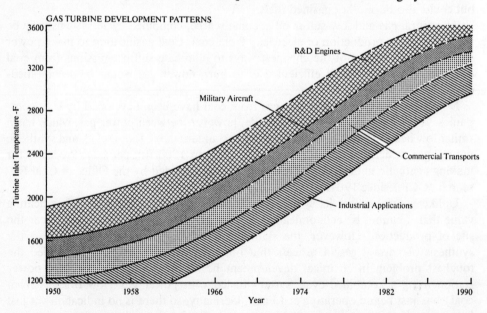

FIGURE 7. The operating temperatures of gas turbines. They have been rising steadily for the last two decades and more improvements are expected. Turbines that can operate with very high inlet temperatures are more efficient than those with lower temperatures. Improvements in military aircraft engines have been reflected in civilian aircraft turbines after about 2 years and in gas turbines for the power industry about 5 years later. [*Source: United Aircraft Research Laboratories.*]

One proposal for a topping cycle is a potassium vapor cycle. Liquid potassium would be heated in the primary boiler to form potassium vapor to drive a gas turbine. As proposed by Arthur J. Fraas of the Oak Ridge National Laboratory, exhaust gases from the fuel burned to heat the potassium would power another gas turbine and waste heat from the potassium turbine would power a steam turbine. A study of the system assuming a particular fluidized-bed process for combustion of coal estimates it will have an efficiency of 51 percent.

The heat-transfer medium for a bottoming cycle could be a fluid, such as isobutane, with a much lower boiling point than steam. A prototype of an isobutane turbine is being developed for production of electricity from geothermal sources.

Impending natural gas and oil shortages mean that new fuels must be found if combined-cycle systems are to play a role in power generation later in this century. Most observers expect that it will be possible to produce power gas from coal, but neither the best method nor the cost is yet clear. It is also possible to produce power gas from residual fuel oil (the relatively inexpensive dregs of the refining process) by methods that are already developed and available for license.

Residual fuel oil with a low sulfur content—usually no more than 0.5 percent by weight—has now replaced coal for power generation in most states east of the Mississippi River. Demand for low-sulfur residual fuel oil is growing and its price is rising. Residual fuel oil with a high sulfur content (2.6 percent) is readily available from Venezuela, and is considerably cheaper (45 cents per million Btu versus 75 cents per million Btu or more). By introducing residual fuel oil into a mixture of steam and air at high pressure, the oil can be partially oxidized to produce a raw fuel gas consisting mostly of carbon monoxide, hydrogen, and nitrogen. The sulfur in the oil is converted into hydrogen sulfide, which can be removed by passing the gas through an absorption system after it has been scrubbed with water to remove carbon and soot particles. Sulfur compounds would be processed to retrieve elemental sulfur and the desulfurized gas would then be burned in the primary boiler of a combined-cycle system. For coal the process would be similar, but complicated by problems of handling a sticky solid which burns to form a troublesome ash.

The technology for partial oxidation of liquid hydrocarbons was developed for making synthesis gas, especially ammonia. It originated in the early 1950's from the Texaco Development Corporation in the United States and the Shell Internationale Petroleum Maatschappi N.V. in the Netherlands. Now both Texaco and Shell have processes for producing low-Btu power gas from residual fuel oil.

A recent evaluation of the Shell gasification process concludes that the unit power cost of electricity in a combined-cycle plant fueled with power gas from residual oil would be about 9 mills per kilowatt-hour, assuming the improvements of turbine temperatures projected for the mid-1970's and assuming the cost of oil to be $2 per barrel. This is higher than current electricity rates (7 or 8 mills per kwhr), but current costs will certainly rise as low-sulfur fuel sources become scarcer.

The chief advantage of gasification technology for electrical power generation is that sulfur can be removed from high-sulfur fuels before they are burned in power stations. The technology to remove sulfur compounds from gas has been well developed by the natural gas industry, which needed to eliminate corrosive gases to protect the pipelines used to carry natural gas. Two of the systems used for natural gas would work well for power gas. One uses hot potassium carbonate and

ENERGY FROM FOSSIL FUELS

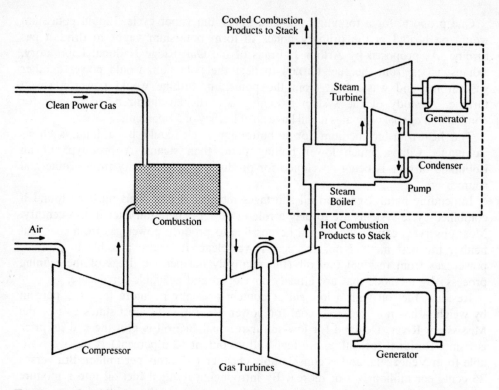

FIGURE 8. Combined-cycle gas and steam turbines fueled by power gas. Air from the compressor is used to make power gas and to burn it in the combustion chamber. Heat in the gases leaving the gas turbine is captured in a steam boiler to produce additional power with the result of high overall efficiency.

[*Source: From "Clean Power from Fossil Fuels," by Arthur M. Squires. Copyright © by Scientific American, Inc. All rights reserved.*]

the other uses various amines as the chemical solvent. In these "wet" processes, the solvent reacts with the carbon dioxide and hydrogen sulfide to form complex compounds, which are retained until the temperature and pressure are changed and then released as the complex compounds decompose. After release, the hydrogen sulfide is selectively oxidized to form elemental sulfur by a system invented by C. F. Claus in 1880. The Claus system is particularly attractive because elemental sulfur is a by-product that power stations can readily market and easily stockpile.

Processes for removing hydrogen sulfide from power gas are expected to be so highly effective (98 to 99 percent removal) that sulfur emissions from a power plant would be almost eliminated. The Shell study estimates that the sulfur concentration in the power gas would be only 5 parts per million, far less than the current limit on stack emissions and less than most estimates for future emission standards.

For removing sulfur from dirty fossil fuels, gasification appears to be superior to the major alternative—removing sulfur dioxide from the stack gases of a conventional power station. Much research money has been spent on the problem of sulfur dioxide removal, but the techniques available still have major problems. Furthermore, gasification may be cheaper. According to Arthur Squires, of the City College of the City University of New York, hydrogen sulfide can be removed from power gas for about $20 per kw, whereas it costs about $70 per kw to remove sulfur dioxide from stack gases.

According to Shell, the gasification process will also produce low nitrogen oxide emission because the nitrogen from the fuel forms N_2 most of the time, and with careful gas turbine design little nitrogen from the air forms oxides. Some observers think a combined-cycle plant could reduce nitrogen oxide emissions by two orders of magnitude, but this has not been demonstrated. Others are more skeptical.

The combined-cycle plant would have absolutely no dust or soot emissions, and advanced combined-cycle designs would reduce thermal pollution by 25 percent compared to conventional power stations. At a reduced level of thermal pollution, cooling stacks would be far cheaper, and it is possible that dumping of hot effluents into natural waterways would not be necessary. Disposal of ash (which would be 50 percent vanadium if Venezuelan residual oil is used) and unburned carbon would be a problem. Shell estimates the carbon waste would be less than 3 percent of the fuel input. Less than 1 percent is highly desirable.

Many consultants, such as Squires and the research team at United Aircraft Research Laboratories, urge immediate installation of combined-cycle systems fueled by low-Btu power gas from residual fuel oil, but an effective gasification process for coal is much more important for the future. The basic chemistry of coal gasification is very similar to gasification of residual fuel oils: partial combustion at high pressures with air and steam. However, coal is much more difficult to handle and transport because it is a solid. Furthermore, the composition of coal varies widely in different parts of the country. A process that works satisfactorily for coal from the East Coast may not work for coal from Illinois or Montana. Another major problem is separating the ash from the coal and disposing of it.

Very little research on gasification of coal for low-Btu fuel has been done in the United States; but for the production of high-Btu gas from coal four well-defined processes exist. By substituting air for oxygen in these four processes, low-Btu power gas can be made. However, the best process for high-Btu pipeline gas may not be the most suitable for power gas (a large methane yield is desirable for pipeline gas but not for power gas).

Because there are many possible methods for each step in low-Btu gasification—handling coal, gasification, and removing ash—a great number of complete processes is imaginable. Few of them, however, have been studied. The only commercially available system was developed in Germany in the 1930's. It is a gravitating bed gasifier, manufactured by Lurgi Gesellshaft für Mineralöltechnik GmbH of West Germany. The Lurgi system has several important limitations. The gasification unit is too small (more than 20 Lurgi gasifiers would be needed for a

Table 3. Composition and heating value of crude power gas from the Lurgi system compared with power gas from a proposed fluidized-bed system. Power gases have a very low heating value compared with natural gas, which is more than 95 percent methane and has a heating value of about 1000 Btu per cubic foot. The fluidized-bed gas has less water vapor, is easier to clean of sulfides because of the low carbon dioxide content, and has a higher heating value.
[Source: Arthur M. Squires, City College of the City University of New York]

Composition (percent by volume)	Lurgi gasifier	Proposed fluidized-bed gasifier
Methane	4.4	.5
Carbon monoxide	10.7	31.8
Hydrogen	15.7	15.6
Carbon dioxide	10.7	.5
Water vapor	27.8	.5
Nitrogen	30.2	50.4
Hydrogen sulfide	.5	.7
Heating value (Btu per cubic foot)	129	157

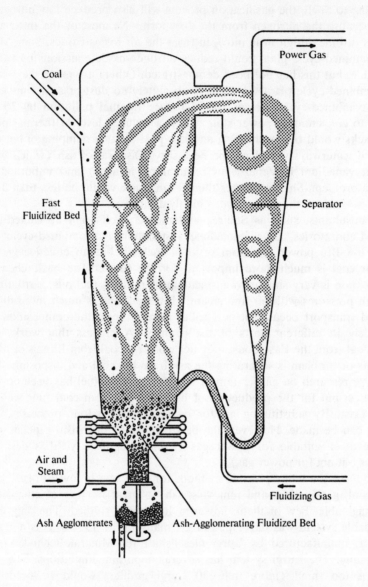

Power Gas

Coal

Fast
Fluidized Bed

Separator

Air and
Steam

Fluidizing Gas

Ash Agglomerates

Ash-Agglomerating Fluidized Bed

FIGURE 9. One proposed method for gasifying coal to produce power gas. This system for partially oxidizing coal at high pressures incorporates a fluidized bed near the bottom where coal is buoyed up by incoming air and steam. Fine particles form a "fast" fluidized bed in the gases rising at high velocity, and the ashes would agglomerate and fall out the bottom. [*Source: Arthur M. Squires, City College of the City University of New York.*]

1000-Mw power station); coal particles smaller than 3 millimeters in diameter cannot be used; and burning Lurgi gas produces large amounts of water vapor.

However, Lurgi gasifiers are available now. A power plant being installed at Lünen, Germany will supply power gas to a combined-cycle system with a gas turbine that will generate 74 Mw of electricity and a steam turbine that will generate 98 Mw. In the United States, the Commonwealth Edison Company of Chicago plans to gain experience with coal gasification by installing three Lurgi units, but will not use the power gas to fuel a combined-cycle system.

The possibility of using coal gasification to make low-Btu power gas has been heavily publicized in the United States in the last year, and the OCR is now considering proposals for its development. But the first improvement on the Lurgi design will almost certainly come from Europe, possibly in Paris, where Albert Godel is planning to test a new design for a gasifier. Godel and Babcock-Atlantique developed a coal gasifier called the Ignifluid boiler 17 years ago, which has been widely used, except in the United States. The Ignifluid boiler successfully makes low-Btu gas with a fluidized bed, a method for burning coal that has several advantages over a fixed bed, such as Lurgi's. Air and steam are injected rapidly to buoy up the granular material and any size coal can be used. Because this fluidization allows easy movement of the solids from one part of the boiler to another, the temperature of the fluidized bed is uniform. Almost no steam appears in the power gas. However, because of the design of the grating on which coal rests, the Ignifluid boiler cannot be operated at high pressures. Godel's new design has a grating that will allow high-pressure operation.

Other proposals for fluidized-bed processes have been studied on paper or with small bench-scale experiments, but have never been tested. Arthur Squires and his colleagues at City College have proposed an elaboration of Godel's design that would treat fine carbon particles and ash differently. In the City College gasifier, the fine carbon particles would form a turbulent fluidized bed (called a "fast" fluidized bed) in the high-velocity gases rising from the fluidized bed composed of larger coal sizes. To collect the coal ash, the gasifier would incorporate Godel's ingenious discovery that coal ash agglomerates at high temperatures in a fluidized bed, which was used in the Ignifluid boiler. Squires and his colleagues are also studying ways to clean hydrogen sulfide from power gas at high temperatures. The methods discussed earlier for cleaning power gas from residual fuel oil could be used, but require that the gas be cooled.

Coal gasification could be very important for supplying energy in the future. Coal may be the only fuel available after gas and oil are depleted, especially if the output of nuclear power stations is limited by technical or environmental problems. But the total amount of money spent on coal gasification research (about $40 million in the last 11 years) is about one-tenth the amount spent on fission every year.

Through various turns of history, economics, and politics, coal research has been badly neglected in the United States. OCR's program for production of high-Btu gas has not yet produced a working pilot plant, and research on low-Btu gas production was not funded at all until $3 million was provided for initial studies in 1972. The shortage of natural gas is here, and the shortage of low-sulfur fuel is imminent. Coal gasification techniques to replace these fuels are needed, but so far the options are not available.

4.

MAGNETOHYDRODYNAMIC POWER
More Efficient Use of Coal

Coal is the largest source of energy now available in the United States, but research to improve the use of coal as a fuel has languished for lack of support. In recent years, however, more attention has been given to cleaning up coal-burning plants because, despite their disadvantages, they are likely to continue as the mainstay of the utility industry for the rest of the century. The magnetohydrodynamic (MHD) generator, which converts heat from combustion gases directly into electricity, is one promising development. MHD generators would require less fuel and produce much less thermal pollution. They may also offer one of the best methods of eliminating sulfur oxide and reducing nitrogen oxide emissions from coal-fired power plants.

Despite its promise, MHD has yet to be demonstrated as a practical technology, in part because support for construction of large-scale experimental facilities has not been available. Substantial technical problems associated with endurance of the equipment remain to be resolved—MHD generators operate at temperatures above 2000°C and the hot residues from coal combustion are extremely corrosive—but most scientists in the field are confident that these will not present serious obstacles. Economic prospects are still uncertain because the technology has not been demonstrated, but preliminary estimates are favorable.

Research on MHD is becoming widespread, with active efforts in Japan and several European countries. Several laboratories in this country are working on MHD with support from the U.S. Department of Interior and the utility industry. The U.S.S.R. is undertaking a more ambitious effort and is already testing an experimental 75-megawatt power plant incorporating an MHD generator fueled with natural gas. Using natural gas makes design of the generator easier than using ash-laden fuels such as coal. Recent U.S. visitors report that the plant has so far produced up to 4 Mw for brief periods and seems to be operating successfully.

The MHD generator is basically an expansion engine in which hot, partially ionized gases flow down a duct lined with electrodes and surrounded by coils that produce a magnetic field across the duct. Unlike the gas in a turbine, the expanding gas propels only itself. Movement of the electrically conducting gas through the magnetic field generates a current in the gas that is collected at the electrodes. Thus MHD generators are compact, have no moving parts, and can accommodate temperatures and corrosive gases that would destroy conventional turbines. Very high temperatures are necessary to ionize combustion gases; but with the addition of small amounts of potassium or other alkali metals, called seed particles, temperatures of 2000 to 2500°C provide sufficient ionization to allow the process to work.

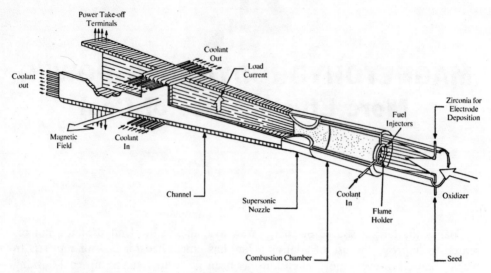

FIGURE 10. One possible configuration for an MHD generator.
[*Source: Richard J. Rosa, Avco Corp.*]

Power plants incorporating MHD generators also include pressurized combustion chambers for burning the fuel and heat exchangers or other equipment for preliminary heating of air fed to the combustor. Preliminary heating appears to be necessary to reach required temperatures, unless oxygen in large quantity is added to the fuel mixture, a procedure which is uneconomical at present. Conventional steam turbines would use MHD exhaust gases to generate additional electricity. In most designs, MHD provides about half the electricity from the combined plant. Overall efficiency of the combined facility is expected to reach about 50 percent, as compared to 40 percent for the best conventional or nuclear power plants; with more sophisticated MHD design, efficiency could reach 60 percent.

MHD generators need stronger magnets than ordinary generators do because of the lower conductivity of gases as compared to copper; superconducting magnets will probably be used in commercial plants. Large superconducting magnets have been built for applications in high-energy physics, but few have been built for MHD purposes and they are still very expensive. Research with field strengths equivalent to those that will probably be used in MHD plants (50,000 gauss) is only beginning. Electricity produced from MHD generators is direct current and must be converted before transmission over existing networks.

Endurance of the generator is the principal problem facing those working on MHD. Only limited experience with long-term operation has been gained—a few-kilowatt generator at the Avco Corporation in Everett, Massachusetts, has been operated for several hundred hours and a 70-kw generator has been run for 500 hours in the U.S.S.R. The major question is whether improved design can prevent leakage of current and arcing between electrodes due to condensation and penetration of the seed material into the generator wall. Other potential problems include plasma instabilities in the ionized gas arising from interactions between the flow and the magnetic field and coal-ash corrosion of the generator walls or of the air heater.

Several generator designs have been proposed to overcome these problems but have not yet been tested. R. J. Rosa of Avco believes that allowing coal ash to condense on the generator walls will build up a protective layer and keep seed material from shorting the electrodes. Research groups at the University of Tennessee Space Institute in Tulahoma, Tennessee, at Stanford University in California, and at the Westinghouse research laboratories in Pittsburgh, Pennsylvania, have proposed still other designs. The Westinghouse team, headed by Stuart Way, proposes hotter wall temperatures to prevent ash buildup; but Way believes that cleaner fuels —such as char, which is produced as a by-product in coal gasification plants—may ultimately be preferable for MHD. Researchers agree that problems can be solved; but long-term testing and further experience with a pilot plant will be necessary to arrive at the best design.

Environmental advantages over traditional coal burning plants may be the most promising aspect of MHD power generation. Sulfur in coal is a major source of air pollution, producing corrosive sulfur oxides when burnt. MHD plants could eliminate these pollutants. Experimental work directed by Daniel Bienstock at the U.S. Bureau of Mines laboratory in Pittsburgh, Pennsylvania, has confirmed that sulfur reacts with the alkali seed particles in the generator to form potassium sulfate or similar compounds. Because the alkali particles are recovered from exhaust gases by cloth traps or electrostatic precipitators—indeed, must be recovered and recycled for economic reasons—essentially all the sulfur can be removed, even from high-sulfur coals.

Nitrogen oxide emissions, which at one time were expected to be a problem because of the high temperatures in MHD plants, also can be reduced, Bienstock finds, if coal is burned in a fuel-rich mixture and excess air is added further downstream. Work at Avco and in Japan has confirmed the Bureau of Mines' results. Detailed calculations at Avco of the kinetics of nitrogen oxide production and con-

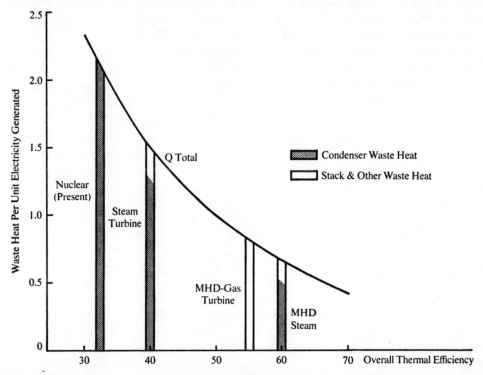

FIGURE 11. Overall thermal efficiency of different electrical generating systems. The waste heat produced increases rapidly with decreasing efficiency.
[*Source: Richard J. Rosa, Avco Corp.*]

sumption indicate that temperature and composition of the combustion gases can be controlled to reduce NO_x concentrations to acceptable levels.

If predicted efficiencies can be achieved, and several independent analyses indicate that they can be, thermal pollution resulting from discharge of waste heat into cooling water would be reduced substantially, and would be less than that of any existing power plant. If gas turbines were used as the second half of an MHD power plant, the steam cycle and the need for cooling water could be eliminated.

Reliable estimates of the cost of MHD plants are not yet available, but there appears to be general agreement that construction costs should be about the same as for traditional coal-fired plants. Operating costs per kilowatt would be significantly lower because of the more efficient use of fuel. Cost of the magnets is the largest single item, and air preheaters are also expected to be expensive. Using fuel enriched with oxygen reduces the need for preliminary heating of air and hence the cost of preheating equipment; but cheap sources of oxygen are not yet available. Hence reductions in the cost of superconducting magnets or the cost of oxygen could further improve the prospects for this source of power. Developing MHD technology will be costly and not without financial risk. In the absence of substantial federal funding, which has not been forthcoming—current R&D spending is about $3 million per year—only limited progress can be made. Both demonstration and commercial power plants will probably have to be large, because the power output of an MHD generator increases in proportion to its volume, while most of the losses increase more slowly in proportion to its surface area; 100 Mw is estimated to be the smallest size feasible for a plant that does not use oxygen enrichment.

Using coal or fuels derived from coal to generate electricity means paying the environmental and human costs of coal mining. But because this energy source is not likely to be replaced—indeed the importance of energy from coal is expected to increase in coming years—efforts to improve the use of this fuel would seem well worthwhile. MHD generating plants would, at the very least, diminish the impact of mining by producing equivalent amounts of power with less fuel. What appears to be needed, as one proponent put it, is to "give MHD a chance."

II.

NUCLEAR ENERGY

"We nuclear people have made a Faustian compact with society: we offer ... an inexhaustible energy source ... tainted with potential side effects that, if uncontrolled, could spell disaster."

—Alvin Weinberg,
Director, Oak Ridge National Laboratory

5.
NUCLEAR ENERGY

Man's first use of nuclear energy was for weapons. But the Manhattan project that led to the development of the atomic bomb during World War II also resulted in the development of several nuclear reactors to manufacture plutonium. Even during the war the possibility that such reactors could provide a new and potent source of energy was foreseen. The hope that nuclear power might prove a major benefit of the peaceful atom led in the late 1940's to an expanded research program under the aegis of the Atomic Energy Commission. Despite the optimism of early proponents, it took many years of effort and large sums of money to develop, first, small power plants for nuclear submarines, and later, large power reactors. Today, nuclear power plants generate only about 4 percent of the electricity used in the United States. Nonetheless, nuclear energy has emerged as a major addition to traditional energy sources and the only new energy technology developed to the point of commercial use in the past 30 years.

Uranium (and potentially thorium) are the raw materials of nuclear energy. One of the unique properties of nuclear fuels is their extraordinary compactness. The uranium fuels used in existing reactors release about 20,000 times as much heat as can be obtained from the equivalent weight of coal; in the more efficient breeder reactors now being developed this ratio will be as high as 1,500,000 to one. The other unique property of nuclear fuels, of course, is their radioactivity. Remote handling of the fuel, special safety precautions, and more complex equipment than that needed for burning fossil fuels are characteristic of nuclear reactors.

Exploring for and mining uranium has become a substantial business. The uranium found in nature is mostly ^{238}U and contains only about 0.7 percent ^{235}U, the isotope that undergoes fission releasing heat within a reactor. For the types of reactors in use in the United States, the fuel must be enriched by increasing the concentration of ^{235}U. The Atomic Energy Commission operates three large gaseous diffusion plants for this purpose, but intends that private industry will build additional plants when the growth of nuclear power outstretches the capacity of the existing enrichment facilities. In addition to mining and enrichment, the rise of nuclear power has created industries devoted to manufacturing nuclear power equipment and fabricating and reprocessing nuclear fuels. By 1980, investment in nuclear power is expected to amount to $60 billion.

The AEC retains considerable control over the nuclear power industry in that it sets standards for and licenses all nuclear facilities. In recent years these standards, particularly those affecting the release of radioactivity from power plants, have been tightened considerably. The AEC's dual role as promoter and regulator of nuclear energy has been challenged, however, and opposition to nuclear power plants on environmental grounds has led to bitter disputes over their siting and delays in their construction. Within the scientific community there has been growing debate over safety features in existing reactors and over the advisability of committing the United States to long-term dependence on nuclear power.

Both in the United States and abroad, research on nuclear power is concentrated on the breeder reactors that will provide heat for the next generation of nuclear plants. The U.S.S.R., Great Britain, France, Germany and Japan all plan to test large breeder reactors between now and 1980. The U.S. demonstration plant will be built near Oak Ridge, Tennessee, beginning in 1974. Federal funding for research on breeder technology will eventually amount to more than $4 billion. Breeders represent the future of nuclear power and what President Nixon has described as "our best hope today for meeting the nation's growing demand for economical clean power." But they also entail a complex new technology and hazards which, in the view of some critics, ought to make the breeder "the last choice of a desperate nation." As this difference of opinion makes clear, the debate over nuclear power is far from over.

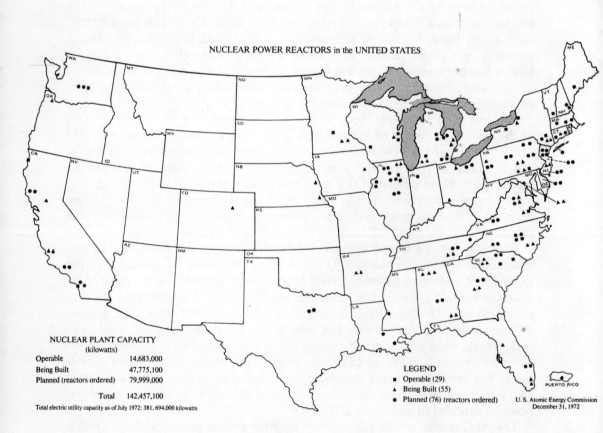

NUCLEAR POWER REACTORS in the UNITED STATES

NUCLEAR PLANT CAPACITY
(kilowatts)

Operable	14,683,000
Being Built	47,775,100
Planned (reactors ordered)	79,999,000
Total	142,457,100

Total electric utility capacity as of July 1972: 381,694,000 kilowatts

LEGEND
■ Operable (29)
▲ Being Built (55)
● Planned (76) (reactors ordered)

U.S. Atomic Energy Commission
December 31, 1972

PUERTO RICO

FIGURE 12. Nuclear power reactors in the United States, present and planned.
[*Source: U.S. Atomic Energy Commission*]

6.

FISSION
The Pro's and Con's of Nuclear Power

Of all the major new sources of energy, nuclear fission has received the most support, and its technology is correspondingly the most well developed. About 30 nuclear power plants have received operating licenses in the United States, another 130 are planned or under construction, and the U.S. Atomic Energy Commission is committed to spending several billion dollars developing the next generation of reactors, the breeders. Some estimates are that nuclear reactors will rapidly become the dominant source of heat for generating electricity, perhaps by the end of the century. Nonetheless, there is growing concern in the scientific community about the extent to which research funds are concentrated on this energy option and about the consequences of large-scale use of nuclear fission as a source of power.

The concerns include reactor operating hazards, particularly the chances of a serious accident, the difficulties of safeguarding fissionable materials used as reactor fuels, and the still unsolved problem of long-term storage for radioactive wastes. Technological failure, earthquakes and other unforeseen natural disasters, and human actions ranging from carelessness to deliberate sabotage are of unusual moment with nuclear power systems. Indeed, the consequences to human health and to the environment of any large release of radioactive substances make nuclear fission potentially the most hazardous of all sources of energy.

At the same time, nuclear fission has substantial advantages over traditional sources of energy. Air pollution from fossil fuels is still a problem, and nuclear power plants do not emit particulates, sulfur oxides, or other combustion products. Nuclear fuel is a more compact source of energy than, for example, coal, and requires less mining. Water pollution, land disruption, and human injuries associated with mining are correspondingly reduced. Transportation costs for nuclear fuels are lower. Fission could also help to replace rapidly depleting fossil fuels as a source of energy and thus conserve them for their value as chemicals.

Nuclear power plants obtain heat energy by splitting atoms of heavy elements such as uranium. But the reactors in commercial service today, primarily light water reactors (LWR's), use nuclear fuel very inefficiently. LWR's utilize less than 1 percent of the energy in naturally occurring uranium—they consume the fissionable ^{235}U isotope and convert only small amounts of the more plentiful ^{238}U into fissionable plutonium. In consequence their fuel supply is limited and is considerably smaller than, for example, known coal reserves. For this reason, LWR's were never considered as more than a stopgap by the early prophets of nuclear power; commercial utilization grew out of the successful effort to develop nuclear power plants

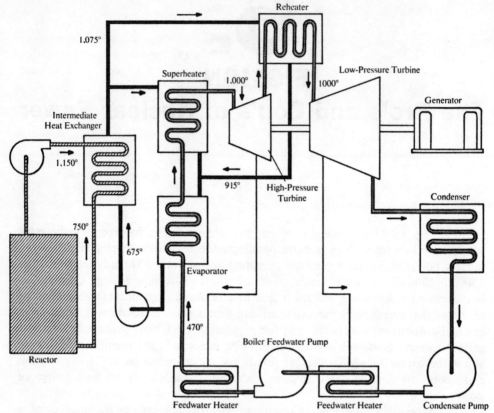

FIGURE 13. Flow plan illustrating the complexity of a liquid metal-cooled fast breeder reactor. The design entails pumping the sodium coolant through the reactor, where it becomes radioactive, and then to an intermediate heat exchanger, where heat is transferred to a separate stream of sodium that is not radioactive. The heat is then used to generate steam, which drives a turbine to generate electricity. Approximate temperatures at various points in the system are given in degrees Fahrenheit.

[*Source: From "Fast Breeder Reactors," by Glenn T. Seaborg and Justin L. Bloom. Copyright © by* Scientific American, *Inc. All rights reserved.*]

for submarines. If fission is to become a major source of energy, breeder reactors will be needed. Breeders produce more fissionable material than they consume and thus theoretically can utilize between 50 and 80 percent of the uranium and thorium resources.

Nuclear power plants based on LWR's are clearly a commercial success. But they have not had entirely the intended impact on the domestic energy picture. Instead of conserving fossil fuels, the expansion of nuclear power has had the perverse effect of stimulating the strip-mining of coal to supply power for the gaseous diffusion plants which concentrate or enrich ^{235}U for nuclear fuels. Conventional generating stations operated by the Tennessee Valley Authority provide most of the electricity for the enriching facilities and are heavily dependent on strip-mined coal.

LWR's have relatively low thermodynamic efficiencies, about 32 percent. Reactors that use helium gas instead of water as a coolant are just now coming into commercial service in the United States. These gas-cooled reactors operate at higher temperatures, with efficiencies of about 39 percent, and also make more efficient use of fuel. Other designs include a reactor with heavy water coolant, which is used in Canada, and a CO_2-cooled model, which is used in Great Britain.

Most nuclear power plants discharge their waste heat directly to lakes or rivers, a practice opposed by environmentalists. Because of their low efficiencies, LWR's produce about 40 percent more waste heat than modern fossil-fired plants. Expected shortages of cooling water in many parts of the country, however, may well force the adoption of systems that discharge heat to the atmosphere. An AEC study estimates that dry cooling systems now under development would raise the cost of power from light water reactors by about 12 percent, but would permit much greater flexibility in siting power plants. If gas-cooled reactors could be operated at high enough temperatures, the helium coolant could be used to drive air-cooled gas turbines directly, eliminating the steam cycle and the need for cooling water altogether.

Periodically, the fuel rods of a reactor must be removed and shipped to a reprocessing plant where fission waste products are removed and unused fuel is refabricated into new rods. Transporting the highly radioactive fuel rods to separate reprocessing facilities raises the possibilities of shipping accidents and theft or loss en route. Reprocessing plants release gaseous radioactive fission products to the atmosphere; these facilities are not now required to meet the strict radiation standards that apply to nuclear power plants. Storing the fission wastes so that the heat of radioactive decay can safely dissipate over long periods of time is also necessary. With the present scale of the nuclear power industry these are not very great problems, although shipping problems and attempts at theft have already occurred, but they promise to become more substantial when nuclear plants proliferate and shipments increase.

Reactor safety, however, has already emerged as a subject of concern. Existing types of reactors have an inherent margin of safety with respect to changes in reactivity in that small increases in temperature within the core have the effect of decreasing reactivity—a negative feedback process in the kinetics of the nuclear reactions which helps to control the reactor. Problems due to failures in the cooling system are more difficult to control because shutting down the reactor does not stop radioactive heating from fission products. Questions about the safety of LWR's have been raised, especially in regard to the still untested adequacy of the emergency core-cooling system. In addition, the discovery of fuel elements damaged by unknown causes in several operating reactors has led many observers to call for a moratorium on increasing the size and power ratings of nuclear plants, both of which have been escalating rapidly.

In West Germany, a government advisory committee on reactor safety has recommended a moratorium on the licensing of LWR's pending further investigation. In the United States, the committee's counterpart has repeatedly expressed concern that reactor pressure vessels could rupture and has urged the AEC in vain to consider such an accident in its safety research. These developments and a series of malfunctions in control systems, safety valves, and other equipment in the relatively few reactors now operating, have led Henry Kendall of the Massachusetts Institute of Technology and other members of the Union of Concerned Scientists—a Boston group of scientists, engineers, and economists—to claim that the chances of a major reactor accident are by no means insignificant.

Light water reactors may continue to be the mainstay of nuclear power for several decades, but breeders—in particular the liquid metal-cooled fast breeder reactor (LMFBR)—are being developed by the U.S.S.R., Japan, several European countries, and the United States. Breeders will eventually eliminate the need for gaseous diffusion plants and will ensure a much larger supply of nuclear fuel, if they prove successful. But this new technology will intensify many of the problems associated with use of LWR's and will raise new ones.

The LMFBR differs from LWR's in that it uses unmoderated, high-energy neutrons and operates at higher power densities—the rate of production of heat per unit volume of the reactor core will be about twice that of LWR's. The higher neutron fluxes within the LMFBR can damage the structure and create materials problems that have not been entirely resolved. Experiments at several laboratories recently showed that stainless steel of the type planned for use in breeder reactors swells when exposed to neutron fluxes for long periods, necessitating changes in the core design to allow for its expansion. Special safety precautions must also be observed for the liquid sodium coolant. Liquid sodium is very reactive with air and water and becomes radioactive, complicating the design and maintenance of an operational plant.

Safety problems of LMFBR's may turn out to be more severe than those of LWR's. The smaller fraction of delayed neutrons from the plutonium fuel in an LMFBR compared to that from ^{235}U in an LWR makes control of the breeder reactor a more delicate process; in essence, the reactor operates much closer to the limits of an uncontrolled chain reaction than LWR's do. Vapor bubbles or other voids in the sodium coolant increase the reactivity within the reactor and special consideration must be given in the design to ensuring that these increases will be self-controlling. Chances of a major loss-of-coolant accident in an LMFBR are expected to be less than in light water reactors; but should an accident occur, the radioactive heating would almost certainly melt the reactor fuel and release large quantities of fission products.

Although LMFBR's will undoubtedly be continually improved, experience up to now illustrates the complexities of the technology. One of the AEC's first pilot breeder reactors suffered a partial melting of its fuel during the course of an experiment in 1955, and the reactor was destroyed. A similar but fortunately much less serious breeder accident occurred in 1966 during the first attempt to operate a commercial power plant incorporating an LMFBR.

Although LMFBR's are the focus of most development work in this country and abroad, gas-cooled breeder reactors and a reactor that uses molten uranium salts as both fuel and coolant have also been studied. Proponents claim that these alternative breeders have a number of theoretical advantages over the liquid metal-cooled breeder, as well as some disadvantages. But they suffer as well from being less developed and hence longer-range options—a disadvantage in the current atmosphere of urgency that pervades U.S. reactor development policy. The LMFBR program consumes almost half of all federal research expenditures on energy technologies and is accorded the highest priority by President Nixon. Construction of a demonstration power plant with a liquid metal-cooled breeder reactor is expected to begin in 1974. One AEC study projects that more than 500 breeder reactors will be in operation by the end of the century, with possibly 2000 such reactors by the year 2020.

The prospect of a nuclear power industry of that size has caused even strong proponents of the breeder to have some second thoughts. Alvin Weinberg, director of the AEC's Oak Ridge National Laboratory, points out that, with the anticipated scale of endeavor, safety factors will have to be improved, decontamination of old reactor sites will be a substantial problem, and new siting arrangements—such as grouping power plants and fuel reprocessing facilities into nuclear parks—may be necessary to minimize the transportation of radioactive materials. Weinberg raises the question of whether social institutions can cope with the demands of large-scale use of nuclear fission, demands that include long-term management of wastes, very high quality manufacturing and construction standards, and unusual skill and vigilance in operating nuclear power plants. Weinberg answers the question in the affirmative; others are less certain.

Reactor wastes, for example, will be a major problem. As much as 27 billion curies of radioactive waste materials may accumulate by the year 2000. These materials will have to be stored for thousands of years under essentially permanent surveillance—implying a permanence to the social order and a degree of responsibility well beyond that of any earlier society.

The properties of ^{239}Pu, the principal fissionable isotope that is produced from ^{238}U within a breeder reactor, also cause concern. A kilogram of plutonium can produce as much energy as about 3 million kg of coal, thus making it uniquely valuable as a fuel. About 1000 kg of plutonium will be required to fuel a large LMFBR. If AEC estimates of the breeding ability of these reactors are correct, the plutonium inventory will double about every 10 or 15 years, and as much as 80,000 kg a year may be produced in LMFBR's by the end of the century.

Handling these quantities of plutonium will present equally unique problems, however, because plutonium is also among the most toxic substances known. Microgram amounts cause lung cancer in experimental animals and federal health standards limit human exposure to a total body burden of 0.6 microgram. The radioactive half-life of ^{239}Pu is 24,400 years, making possible the essentially permanent contamination of an area in the event of a major accident. Plutonium has a critical mass of about 5 kg—a relatively small amount compared with the contemplated inventory of a large reactor or fuel-processing facility.

The AEC takes the position that despite the dangers of plutonium and the so-far undistinguished record of the agency and its contractors in handling and safeguarding this material its widespread use will present no unresolvable problems. But the real consequences seem to depend on how safely breeder reactors can operate in practice and on how well safeguards now being developed will work.

Technological problems can in theory be solved, but the social problems arising from the misuse of technology are less easily dealt with. In particular, the vulnerability of nuclear power plants and fuel processing facilities to sabotage and the potential for diversion of plutonium to illegal purposes could lead to difficult situations. The AEC is concerned that the value of plutonium, about $10,000 per kg, may be an incentive for the creation of a black market, and the 500 shipments of plutonium per week expected by the end of the century would offer ample opportunity for hijacking. The information necessary to construct a crude nuclear bomb is readily available and despite the fact that reactor-bred plutonium does not have the ideal isotopic composition for weapons, extremist groups might find such a project tempting. Even without bombs, the hazards to public health and national security of plutonium diversion would be substantial.

Breeder reactors do have several advantages over the current generation of reactors. They will, for example, have thermal efficiencies approaching 40 percent. More significantly, cost of an LMFBR is less tied to the cost of uranium, so that breeder reactors theoretically could use even very low grade ores, should that prove necessary.

Spokesmen for the AEC have in the past advanced several reasons for the rapid development of the breeder technology, including a projected shortage of uranium to fuel LWR's and the economic advantages of the breeder. There is no question but that breeder reactors will be needed eventually if fission is to continue to be an economic means of generating power; but there is considerable disagreement with the AEC's admittedly conservative estimates of uranium reserves and with their optimistic estimates of how soon breeders might be economically competitive. At issue is the wisdom of rushing the development of a difficult technology and hastening the accumulation of large plutonium inventories. A growing number of scientists believe that research on the breeder ought certainly to be continued and even broadened, but that the commercialization of this technology could well be delayed a number of years—long enough to ascertain whether other, and less hazardous, sources of energy can be made available.

The use of nuclear power thus poses a considerable dilemma. Fission can become a major energy source, probably the only large energy source, other than fossil fuel, for which the technology is now reasonably assured. But its widespread use may prove a mixed blessing.

7.

BREEDER REACTORS
The Future of Nuclear Power

Current efforts to develop breeder reactors, if successful, will markedly affect the structure of the U.S. power industry. Growing demand for electricity and projected shortages of high-grade uranium ores make it likely that breeder reactors will eventually generate a substantial part of the country's electricity. Breeder reactors will offer, among other advantages, lower thermal pollution and more efficient use of uranium reserves than conventional light water nuclear power plants.

Prototype generating stations powered by breeders, so called because the reactors produce more fuel than they consume, are nearing completion in France and Britain. A prototype in the U.S.S.R. is already in operation. Ambitious programs to develop breeder reactors are also under way in Germany, Italy, and Japan. In 1971, President Nixon announced long-delayed U.S. plans to build a demonstration plant, and has more recently indicated his support for a second such plant. But construction of the first plant—which could take 6 to 7 years—appears unlikely to start before 1974. Prospects for viable, commercial-sized plants are even more uncertain, although the announced intention of the AEC is to achieve this goal by the mid-1980's.

The attractiveness of the breeder reactor is that it can utilize thorium and the more abundant form of uranium (^{238}U); thus, its fuel costs might be substantially lower than those of conventional nuclear reactors. This advantage will be enhanced if, as projected, demand for electricity doubles more than twice before the end of the century and the cost of uranium rises as easily accessible ores are exhausted.

Energy released in breeder reactors, as in other nuclear reactors, comes from the fissioning of uranium or plutonium atoms. In fission, more than two neutrons are emitted (on the average) for each atom fissioned. Neutrons not needed to maintain the chain reaction are used in breeder reactors to convert "fertile" isotopes of the heavy elements into new supplies of fissionable fuel. A similar conversion occurs less efficiently in the light water reactors in commercial use today. But even if the plutonium they produce is recycled as fuel, these reactors can use only about 2 percent of the uranium that is mined. In contrast, breeder reactors, because of their greater neutron economy, may be able to use 50 to 70 percent, a saving that the AEC expects will reduce uranium needs by 1.2 million tons over the next 50 years.

Breeders burn either uranium or plutonium, and at the same time convert thorium (^{232}Th) or ^{238}U into, respectively, ^{233}U or plutonium (^{239}Pu), which are fissionable materials. In the conversion process, a neutron is captured by the nucleus of a fertile atom and beta particles (electrons) are released. The ^{232}Th-^{233}U cycle is feasible in a so-called thermal reactor, in which neutrons are slowed by collisions

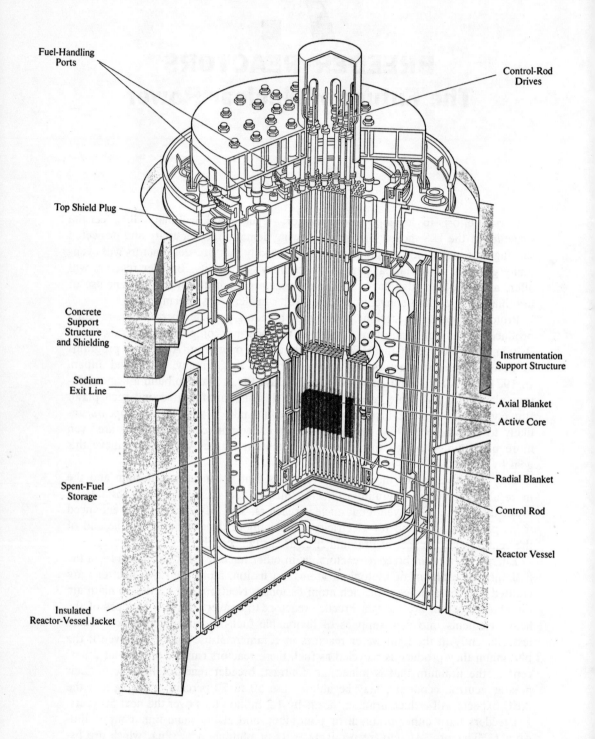

Fuel-Handling Ports

Control-Rod Drives

Top Shield Plug

Concrete Support Structure and Shielding

Sodium Exit Line

Instrumentation Support Structure

Axial Blanket

Active Core

Spent-Fuel Storage

Radial Blanket

Control Rod

Reactor Vessel

Insulated Reactor-Vessel Jacket

FIGURE 14. Fast breeder reactor of the liquid metal-cooled type that will be incorporated in a demonstration power plant to be built near Oak Ridge, Tennessee.
[*Source: From "Fast Breeder Reactors," by Glenn T. Seaborg and Justin L. Bloom. Copyright ©* by Scientific American, *Inc. All rights reserved.*]

with a moderator (usually water or graphite) to energies of about 100 electron volts. The more efficient ^{238}U-^{239}Pu cycle can utilize unmoderated neutrons with energies thousands of times higher, and reactors based on this concept are known as fast reactors. The absence of a moderator in a fast reactor has the advantages that fewer neutrons are unproductively absorbed—so that new fuel is created at a higher rate—and that the reactor core can be considerably smaller.

The more efficient a reactor's use of neutrons, in theory, the lower the cost of the power it produces. A measure of neutron efficiency is a breeder's doubling time —the period required to produce roughly twice the fissionable material originally present. Theoretical doubling times of various proposed breeder reactors range from 6 to 20 years, although some nuclear experts believe that doubling times greater than about 10 years will not be economical. Other factors, such as the costs of fuel fabrication and reprocessing, safety features, and plant operation and construction also help determine whether a reactor will be economical.

The primary effort, both in the United States and elsewhere, is aimed at developing a fast breeder design that will use liquid sodium as a coolant and heat transfer medium and plutonium oxide as fuel. But this goal, and how it is being pursued, has aroused considerable disagreement within the U.S. nuclear community. Considerable effort has been expended in the U.S. program to develop reliable heat exchangers, pumps, valves, and other hardware for handling the highly reactive sodium. According to some critics, however, current designs are so conservative that they may well be uneconomical. Others are disturbed by the pace of the U.S. effort, which costs more than, for example, Britain's, but seems to be progressing more slowly. Research on more economic fuels and alternative reactor concepts has been so cut back that it now receives only token support.

Nonetheless, development of the liquid metal-cooled fast breeder reactor is the main hope for the near future. Because the sodium passing through the reactor core becomes very radioactive, an intermediate heat exchanger is necessary. The primary coolant "loop" transports the liquid metal through the reactor core. The heat is then transferred to a second, independent loop of nonradioactive sodium, which flows through the steam generator. The sodium must be kept extremely pure because even small amounts of impurities can obstruct the sodium flow and can cause the mixture to become extremely corrosive, possibly causing leaks in the intricate plumbing.

In current designs, the fuel is formed into pellets about 0.6 centimeter in diameter and about twice as long, which are packed into stainless steel fuel rods. Plutonium oxide, or a mixture of plutonium and uranium oxides, is the fissionable material in the reactor core. The fuel may be diluted somewhat with ^{238}U, and additional fuel rods of this fertile material are located in the blanket around the core.

Breeder fuels must withstand extremely high rates of neutron irradiation for long periods within the reactor—long enough for a fuel "burn-up" corresponding to fission of 10 to 12 percent of their heavy atoms. Only a few experimental fuel rods have been tested under these conditions. (Fuels in the current U.S. power reactors achieve a fuel burn-up between one and three percent before they are reprocessed). Plutonium carbide and other advanced fuels have also been studied and may be preferable to oxide fuels, but less is known about their behavior under heavy irradiation.

The fast neutrons in the core of a breeder reactor also have important effects on the stainless steel used as structural material and as cladding for the fuel rods. Radiation damage in the steel creates small voids that grow and cause the material to swell, become brittle, and finally fracture, an effect that was discovered only a few years ago. No data are available on the amount of swelling that will occur at the neutron fluxes (3×10^{23} neutrons per square centimeter) expected in commercial-sized LMFBR's, but AEC scientists have estimated that volume changes as high as 10 percent may occur. The swelling may increase the doubling time of breeders, because additional (unproductive) space must be designed into reactor cores to allow for it, and may raise the costs of operating breeders.

A second design, which has so far received a lower priority, is the gas-cooled fast breeder reactor (GCFBR). In this concept, an inert gas such as helium transfers the reactor's heat to the steam generator. Because the helium does not become radioactive, no intermediate heat exchanger is needed. But because the heat capacity of a gas is smaller than that of a liquid metal, the gas must be compressed to pressures of 70 to 100 atmospheres, and the entire reactor must be enclosed in a pressure vessel.

Early designs for gas-cooled breeders were based on the assumption that extremely high temperatures and advanced fuels would be required and that elaborate safety precautions would have to be taken because of the high pressures involved. The development of prestressed concrete vessels has alleviated some of the safety concerns. These do not undergo sudden failures and, in the event of leaks, have a tendency to self-seal because they are always under compression. And recent studies have indicated that a GCFBR could run on essentially the same fuel and at the same temperature as a liquid metal-cooled breeder. Gas-cooled thermal reactors are already in commercial operation, so that substantial experience with this coolant technology is already available.

Possible advantages of gas-cooled breeders are several. Most important, they are expected to have a shorter doubling time than the LMFBR because helium absorbs fewer neutrons than sodium, and so is less of a moderator. Some estimates indicate a doubling time of less than 10 years, which is more efficient than many scientists expect the initial sodium-cooled plants to be. Helium does not become radioactive and, unlike sodium, cannot react with air and water should a leak occur. Because helium is transparent, maintenance of gas-cooled reactors is expected to be easier. The bubbles that can form in sodium and cause overheating problems cannot occur in a gas-cooled system.

The major disadvantages of the gas-cooled breeder are high operating pressure and the need to maintain forced circulation. In the event of a reactor accident involving the coolant circulating system, for example, the gas-cooled reactor is dependent on mechanical equipment such as blowers to circulate the gas, so extremely reliable back-up equipment is required. Because current gas-cooled reactor designs include a vented fuel element to equalize internal and external pressures, the radioactive fission gases produced in the fuel must be passed through a purification system enclosed within the reactor vessel. Problems caused by swelling of the fuel elements also exist, as in the LMFBR.

Hopes for a commercially viable breeder in the immediate future depend on developing fast breeder reactors, either the LMFBR or the GCFBR. On a longer time scale thermal breeders may also be very attractive. Research on a thermal breeder fueled with molten uranium salts has been carried out at the AEC's Oak Ridge, Tennessee, laboratory. A molten salt breeder would operate at very high temperatures, so that special materials will be necessary; but its compact size and the small amount of fuel required are expected to result in relatively low capital costs and an extremely short doubling time. Because the fuel is molten, continuous, on-line

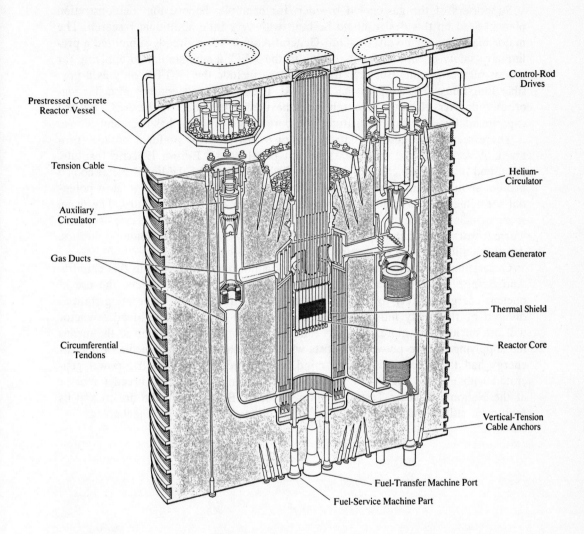

FIGURE 15. Fast breeder reactor of the gas-cooled type. Such a reactor differs from a liquid metal-cooled reactor primarily in that the helium coolant must be highly pressurized to obtain adequate cooling. Because of the high pressures, the reactor is enclosed by a massive prestressed-concrete vessel.

[*Source: From "Fast Breeder Reactors," by Glenn T. Seaborg and Justin L. Bloom. Copyright © by* Scientific American, *Inc. All rights reserved.*]

reprocessing would be possible. All the details of this novel concept have not been proved, but confining power generation and fuel reprocessing to one location would have both environmental and economic advantages.

There appears to be general agreement that the liquid metal-cooled breeder, for which designs are now furthest advanced, should be built. Indeed, President Nixon's announced support for the LMFBR raises its status very nearly to that of a national goal. But some scientists and engineers believe that it is unwise to commit all U.S. resources—and the future of the country's supply of electric power—to what is still an economically unproven system; they believe that alternatives should be vigorously pursued.

Supporters of the gas-cooled breeder, for example, believe that demonstration plants based on this design could be built with very little additional research. The major manufacturer involved—Gulf-General Atomic—has already submitted a preliminary safety document for its design to the AEC, a step similar to applying for a construction permit. But present AEC plans include the GCFBR only as a possible long-range follow-up to the LMFBR. The molten salt concept also has vigorous supporters. Limited research funds have been provided by the AEC, but the experimental reactor that constituted the core of the program has been shut down.

There seems no question that the utility industry would prefer a broader program. A report of the reactor assessment panel of the Edison Electric Institute, published in 1970, concluded that although development of the liquid metal-cooled breeder should continue, the gas-cooled breeder seemed to offer the greatest potential for achieving low cost electric power before the end of the century. The panel recommended that development of the gas-cooled breeder receive greater support. Current work on this concept is supported primarily by a group of some 50 utilities. Another utility group is supporting work on the molten salt breeder.

Generating electricity with nuclear energy is no longer a novelty in this country. And despite concerns about reactor safety and environmental effects, the use of nuclear energy is likely to expand further—more than half the generating stations ordered by the utility industry in 1972 are nuclear-powered. The breeder reactor will use nuclear fuel more efficiently and can be a significant addition to the means for supplying electric power. Scientists who first dreamed of the benefits of nuclear energy had the breeder reactor in mind, and in fact the first electric power generated with nuclear energy, in 1951, came from an experimental breeder reactor at the National Reactor Testing Station in Idaho. But between the dream and its fulfillment still lie substantial steps in the development of this new technology.

III.
ALTERNATIVE ENERGY SOURCES

"A major factor contributing to our present energy crisis is that the necessary research and development efforts which could have provided us with the technological options and capabilities we now need so desperately were not undertaken in the past."

—SENATOR HENRY M. JACKSON

8.

ALTERNATIVE ENERGY SOURCES

Even if further research significantly improves the use of fossil and nuclear fuels, it is nonetheless clear that other energy options are needed. Fossil fuel resources in the United States are running out faster than nuclear fuels can replace them, and the growing reliance on nuclear fuels could be dangerous and environmentally damaging. Many alternative ways to obtain energy are conceivable, and they are all ultimately derived from four natural sources of power: the radiation of the sun, the energy of the tides, the heat inside the earth, and the energy stored in the atomic nucleus more than 10 billion years ago when our galaxy was formed.

The energy radiated by the sun is immense and is constantly being replenished. The power of the tides is also a renewable resource, estimated at 3 billion kilowatts. The rate of heat convection to the surface of the earth from geothermal sources is not well known, but is estimated to be about 300 million kw. Much more power, however, could be extracted from geothermal deposits created during earlier ages. Some fraction of the geothermal resource is renewable, while the rest is depletable. The energy of the atomic nucleus is certainly depletable—fission fuels are getting scarce—but if fusion can be proved the supplies will last a million years or more.

Of the three major alternatives to the proven sources of power, fusion is the only one that is now adequately funded (at about $65 million per year), geothermal research is funded with a few modest programs, and solar research is almost completely neglected. Yet solar radiation is man's greatest resource. Solar energy not only heats the earth's surface, but drives the rains, winds, and ocean currents, and provides the energy for all plant and animal life-cycles through photosynthesis. In units of a billion kw, the total power continually radiated from the sun and intercepted by the earth is 173,000, while the power used by all the industrialized societies of the world is only one. (The United States uses about one-third of a billion kw.) About half the sun's radiation is converted directly to heat at the earth's surface, about one-fourth is reflected into space, and about one-fourth is spent in evaporation and precipitation cycles, which—among other things—provide the energy of currents in streams and rivers. A very small fraction of the earth's total solar radiation drives the atmospheric and oceanic circulations and convections (370 billion kw). An even smaller amount (which is nonetheless 40 times the world's power consumption) is captured by the chlorophyll in plants and sustains photosynthesis.

Harnessing the power of the sun to produce energy for an industrial society is hardly an unproven idea—but so far the sun's energy has only been utilized indirectly. In 1850, for example, almost 90 percent of the energy used in the United

States came from wood, and of course, the energy released in burning wood is just the solar energy that was stored when carbon dioxide was bound into carbohydrates during photosynthesis. The energy of fossil fuels, too, originally came from the sun, building up slowly over about 600 million years as organic materials that did not immediately decay were buried under sediments. Presumably this process is continuing at about the same rate as in the past, but replenishment is so slow it is insignificant. Although fossil fuels are, in fact, renewable, one million years from now only one six-hundredth of the fossil fuel inheritance of industrial civilization will have been replenished. Most of the world's supply will probably have been extracted and burned in the short time of 300 years, and the fossil fuel epoch will prove to be extremely transitory when viewed over the span of human history. For the long run, more quickly renewable energy sources or more abundant ones are needed.

Like fossil fuels, solid organic wastes have an energy content that is ultimately derived from solar energy through photosynthesis. If all solid wastes were converted to synthetic fuel, the available energy would probably be only about 3 percent of the U.S. energy consumption. Nevertheless, conversion of solid wastes to fuel could solve the difficult problem of their disposal, at the same time providing a small but constant supply of fuel.

Hydroelectric power is another well-proven indirect use of solar energy, because the sun's radiation drives the cycle of rains that supplies surface runoff water. Hydroelectric power is renewable, is still the cheapest source of electricity in the United States, and provides more than twice as much electricity as nuclear power plants. The world's total hydroelectric power is estimated to be three billion kw, of which only 8.5 percent is developed. The potential for water power in Africa, South

Table 4. Estimates of depletable energy resources of the United States in units of U.S. annual energy consumption (6.6×10^{19} joules).

Resource	Recoverable	Total
Fossil fuels		
Coal	125	1300
Petroleum	5	280
Natural gas	5	110
Oil shale		2500
Nuclear fission		
Conventional reactors	2.3	15
Breeder technology	115	750
Nuclear fusion		
Deuterium-deuterium		$\sim 10^9$
Deuterium-tritium		$\sim 10^6$
Geothermal heat		
Steam and hot water	0.2	> 60
Hot rock		>600

The figures in the table are equivalent to the number of years that the resource would last, if all energy came from that source. Recoverable resources are defined as those known and *now* available; total resource estimates include expected off shore deposits and do not necessarily represent recoverable amounts of energy.
[Sources: U.S. Geological Survey and Los Alamos Scientific Laboratory]

Table 5. Estimates of renewable energy resources in the United States in units of the current U.S. *rate* of energy consumption (6.6×10^{19} joules per year).

Resource	Continuously available
Solar radiation	740
Wind power	5
Sea thermal gradients	>6
Hydropower	0.14
Photosynthesis	0.23
Organic wastes	0.1
Tidal energy	0.1

The numbers in the table indicate the *proportion* of current U.S. energy needs that the resource could supply for an indefinite period.
[Sources: Chauncey Starr and the National Science Foundation]

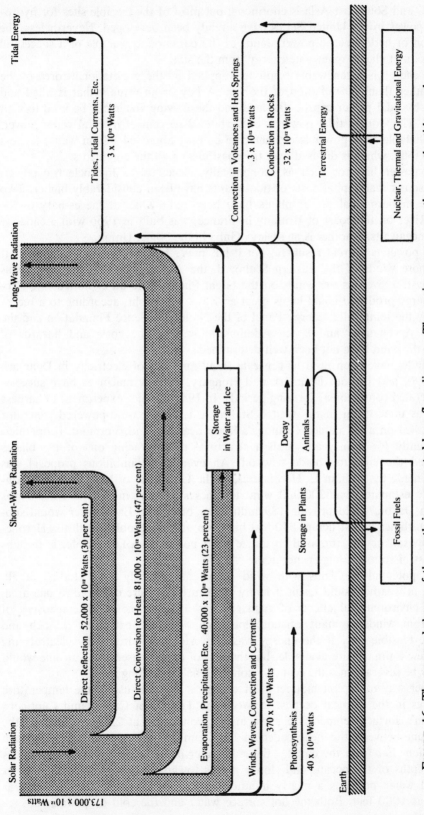

FIGURE 16. The energy balance of the earth is represented by a flow diagram. The major sources are the sun, the tides, and energy from nuclear, thermal and gravitational sources. More than 99 percent of the incoming energy is solar radiation, and the apportionment of solar energy is indicated by the horizontal bands. The smallest portion goes into photosynthesis. Some energy is stored as a result of evaporation and photosynthesis, but most solar energy is either directly reflected from the atmosphere [short-wave radiation] or radiated into space [long-wave radiation]. Tidal energy and geothermal energy inputs are much less than solar. [Source: From "The Energy Resources of the Earth," by M. King Hubert. Copyright © by Scientific American, Inc. All rights reserved.]

America, and Southeast Asia is enormous; but most of the feasible sites for hydro-electric power in the United States have already been developed. Nevertheless, the utilization of hydroelectric power stands as the paramount example of a successful way to extract cheap, renewable energy from the sun.

Tidal energy is a renewable resource, supplied by the gravitational force of the moon rather than the radiation of the sun. A bay or an estuary that is filled and emptied by tidal power can be dammed and the flowing water can be used to turn turbines. The world's tidal power is only about 2 percent of its total water power, and exploitable tidal power is estimated to be only about 64 billion watts, far less than the total, which is diffused over the coastlines of all the continents.

Tidal power has not been as economically successful as hydroelectric power. One reason is that capital costs of tidal plants have been considerably higher. Two full-scale modern tidal power plants have been built. One, at the estuary of the Rance River on the coast of Brittany in France, was built in 1966 with a capacity of 240 megawatts. Another is at Kislaya Guba in the Soviet Union.

Wind power is a great resource, but it has never been used on a large scale. If the offshore winds of the Eastern Seaboard, the Aleutian Islands, and the Texas Gulf Coast, as well as the winds on the Great Plains, were all exploited, the electrical energy produced could be as great as 1.5×10^{12} kwhr, according to a recent report by the joint Solar Energy Panel of the National Science Foundation and the National Aeronautics and Space Administration. But the costs and hazards of utilizing the wind have not been well determined.

Windmills have been used to generate small amounts of electricity in Denmark since 1890, and in both Denmark and Hungary, 200-kw machines have successfully operated continuously for long periods. In 1941 a major experiment to harness winds was undertaken in the United States. A 1.25-Mw wind-powered generator was installed on a hill named Grandpa's Knob near Rutland, Vermont. It operated intermittently for four years, until a structural piece bracing one of the blades broke and the eight-ton blade flew off. An even more ambitious proposal was recently made by William E. Heronemus, of the University of Massachusetts, Amherst. He estimates that 300,000 wind turbines stretching in a band from Texas to North Dakota could provide 189 million kw of power. Each tower would support 20 turbines in a structure 850 feet high. An obvious problem with the Heronemus proposal, or any proposal to utilize wind power on a large scale, is the unsightliness of so many huge generators.

Other potential problems with wind power have hardly been studied at all. Changes in weather could result if many wind turbines were operated in one area, and the environmental effects of such changes are not known. The recovery of power from winds in many offshore locations would be extremely difficult and might be feasible only if the energy was converted to hydrogen by electrolyzing water. Few estimates are available, but the cost of a land-based wind turbine would probably be several times the cost of a conventional generating plant.

Another way to exploit indirect solar energy is to make use of the temperature differences in the tropical oceans. Between the Tropics of Cancer and Capricorn the ocean's surface temperature stays almost constantly at 25°C because of the equilibrium between the heat collected from sunlight and the heat lost through evaporation. Far from the equator the cold water melted from snow and ice slides to the depths of the oceans and slowly moves toward the equator. In the tropics this cold water provides a nearly infinite heat sink at about 5°C at a depth as shallow as 1000 feet. Both the hot surface water and the cold deep water are re-

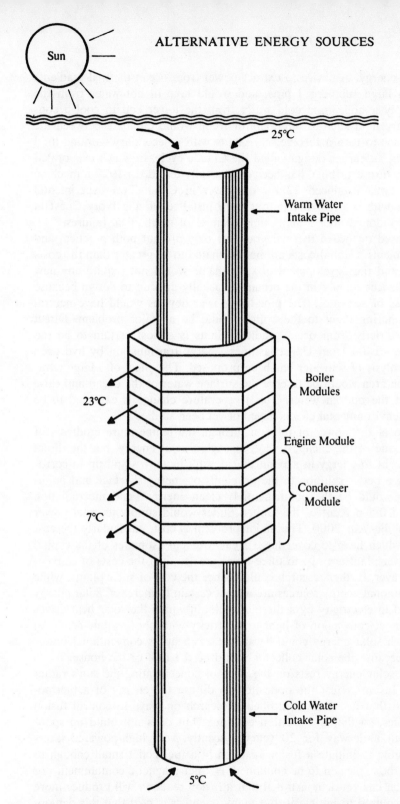

FIGURE 17. An artist's impression of a solar power plant utilizing the thermal gradient of tropical oceans. Water at 25°C would be taken in at the top and water at 5°C would be taken in at the bottom. The difference in temperatures could be used to drive a turbine. The entire device would be neutrally buoyant in several hundred feet of water.

[*Source: From "Solar Sea Power," by Clarence Zener, Physics Today, January 1973.*]

plenished by solar energy. A device to extract power from ocean thermal gradients would look like a large submerged pipe, and would take in hot water from the upper end to supply a boiler and cold water from the lower end to cool a condensor. A secondary fluid such as ammonia or freon would circulate between the boiler and condensor to turn a turbine. No plants with a secondary working fluid have been built, but an earlier design called the Claude cycle, in which evaporated seawater is used to turn a turbine, has been successfully tested. In 1929 a machine using the Claude cycle produced 22 kw of power in Cuba. Two experimental Claude cycle units with 3.5-Mw net output were installed off the Ivory Coast in 1956, but were abandoned after a short time because of mechanical failures.

Power plants based on ocean thermal gradients may present both practical and environmental problems. Capital costs are not estimated to be greater than the costs of nuclear plants, and the development of such plants would not require any new technology—but devices for use in the ocean are usually difficult to design because of the corrosiveness of seawater. The pipe-like power devices would have neutral buoyancy, but anchoring them to the bottom could be a major problem. Direct transmission of electricity from ocean-gradient plants is almost certain to be too costly if the plants are far from U.S. coastlines; energy transmission by hydrogen produced by electrolysis of seawater has been proposed. The effect of a large number of plants in one area would be to cool the surface water of the ocean and raise the temperature of the bottom water. This temperature change is expected to be very slight, but its environmental effects have not yet been studied.

The indirect use of the power of the sun through the temperature gradients of the ocean may be one of the cheapest ways to exploit solar energy, but the direct conversion of sunlight to energy in a plant that doesn't have to stand up to corrosion appears to be a better solution to the problems of energy shortage and pollution. Solar energy would be the quintessential "clean energy," and the radiation falling on less than 0.5 percent of the United States would supply its total power needs expected for the year 2000. The major problem is cost, particularly the cost of the collectors, which have to cover large areas. Current estimates of the capital cost of a solar power plant are two to three times greater than the costs of conventional plants. However, further research could reduce the cost of solar plants, while the costs of conventional energy sources are almost certain to increase. Solar energy could be converted to electricity by a thermal cycle, in which the solar heat drives a turbine, or by direct conversion of light to electricity with photovoltaic cells. An application in which solar energy could be almost as cheap as conventional power is home heating, because the solar collector would be the roof of the house.

The promise of solar energy rests on the hope of concentrating the sun's radiation; the hope of fusion is just the opposite—to diffuse the energy of a thermonuclear reaction so that it can be controlled. Research on ways to control fusion with magnetic forces, so that the reaction continues but does not build up spontaneously, has been underway for 20 years. Recently, very high-powered lasers have made it possible to initiate a fusion reaction in a fuel pellet small enough to allow the force of the explosion to be contained. Neither magnetic containment nor laser fusion research has yet demonstrated that a fusion reaction will produce more energy than that required to initiate it, but many scientists expect that this demonstration will be achieved in the 1980's.

Geothermal power, a resource that has already been proved practical in several areas of the world, is really an indirect form of fission energy. Natural radioactive minerals decaying in the earth's core produce heat that is carried upward to produce steam and hot water near the surface. Geothermal plants in California, Italy, and New Zealand use hot steam for power, and far more energy is available from hot water and hot rock. But estimates of the total resource vary by several orders of magnitude. One of the most important goals for research in geothermal energy is to determine more precisely the size and type of the geothermal resources available.

9.

GEOTHERMAL ENERGY
An Emerging Major Resource

The earth's heat is a potentially valuable source of energy. In the opinion of the university scientists and industrial engineers who have examined its possibilities, this heat could be used to generate substantial amounts of electricity in the near future; geothermal resources, they believe, are large and can be readily exploited. Three types of resources are being considered—steam, hot water, and hot rock.

At the Geysers in northern California, generating plants powered by geothermal steam produce 180 megawatts of electricity at costs lower than those for comparable plants using fossil or nuclear fuels. Plans have been announced for utilizing sources of hot water, a much more abundant resource than steam, to generate electricity and to ease the chronic water shortage in the southwestern part of the United States. Means of tapping the still larger resources of subterranean hot rock have been proposed but have not yet been proved technically feasible, and field experiments to test the concept have begun.

Despite the optimistic outlook, geothermal energy use is in its infancy and substantial technical problems remain to be solved. Very little exploration for deposits of heat has been done; prospecting techniques are in the early stages of development. Controlling corrosion by mineral-laden hot water and designing turbines that can operate efficiently at the low temperatures of many deposits may be crucial to the exploitation of this source of energy on a large scale. Geothermal power plants will not be without potential environmental problems, including air and water pollution and subsidence of land or seismic disturbances caused by pumping, but none of these are regarded as insoluble. Efforts to develop geothermal power have for the most part been confined to industry. Federal support of research has been extremely limited, and indecision in establishing policies to guide the leasing of resources on federal lands has delayed their development.

Geothermal heat has been described as a form of fossil nuclear energy, since it is produced primarily by the decay of radioactive materials within the earth's interior. Radiative and conductive processes transport small amounts of heat to the surface, but large deposits of heat within the earth's crust are apparently the result of geologically recent intrusions of molten rock from the mantle. Where groundwater comes into contact with hot rock, natural deposits of steam or hot water may form. Surface manifestations of these deposits such as hot springs are found in many parts of the world, but they appear to be concentrated in regions of recent volcanism and at the boundaries of the major crustal plates. At sufficient depths, however, hot rock can be found anywhere; in much of the western United States, temperatures of 300°C are estimated to occur within 6000 meters of the surface.

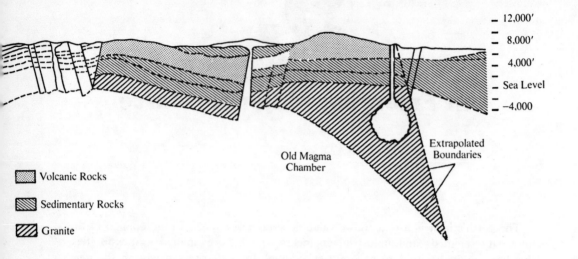

FIGURE 18. Proposed geothermal reservoir at the edge of the Jemez Caldera, an extinct volcano near Los Alamos, New Mexico.
[*Source: Los Alamos Scientific Laboratory.*]

Electric power is being produced commercially from geothermal energy in seven countries, including New Zealand, Japan, and the Soviet Union. Steam from the region around Larderello, Italy, has been used to generate electricity since 1904. In the United States, the only geothermal resource in commercial use is the Geysers steam field north of San Francisco. Most of the easily accessible deposits of heat in the United States are believed to lie in the West, although little prospecting in the eastern part of the country has been done.

At the Geysers, a group of companies headed by the Union Oil Company drill for and produce steam and sell it to Pacific Gas and Electric (PG & E), which uses it to generate electricity. Steam from the wells is collected, filtered to remove abrasive particles, and passed through turbines. The steam has relatively low pressure and temperature—typically 100 pounds per square inch and 205°C at the Geysers compared to 3000 psi and 550°C for the steam in some modern fossil-fueled power plants—and the turbines therefore have a different design. Exhaust steam from the turbines is condensed, and the resulting water is used in cooling towers. Most of the water is evaporated to the atmosphere in these towers, conveying with it the waste heat from the power plant. About 20 percent of the condensed water, containing trace chemicals such as boron and ammonia, which would pollute local streams if released, is reinjected into the ground through deep wells.

Because of the lower pressures and temperatures at which they operate, the turbines at the Geysers are about a third less efficient than those of conventional power plants and require about 450 Mw of heat to produce 100 Mw of electricity. Nonetheless the Geysers plants cost less to build and operate than comparable fossil-fueled or nuclear power plants in California, making geothermal energy PG & E's least expensive new source of electricity. The utility plans to add 110 Mw of generating capacity at the Geysers every year for several years. Estimates of the size of the resources in the Geysers region range from 1000 to 4000 Mw. How soon the supply of steam from a given well will be depleted is not known, and this uncertainty contributes to the cautiousness with which the utility is expanding its use of geothermal power.

Wells in the Geysers field, like those in the Larderello region of Italy, produce steam unaccompanied by liquid water, which makes power generation easier and cheaper. Sources of "dry" steam are rare, however; geothermal wells more commonly produce a mixture of steam and hot water, which must be separated before the steam can be used to generate electricity. In some places, like New Zealand, the water is simply discarded into stream beds. In others, such as those near the Salton Sea in California, the water has a very high content of dissolved minerals and cannot be discharged into the environment. Most proposals for generating power with steam extracted from hot water wells are based on the assumption that the water would be either reinjected into deep wells or treated to remove its minerals, thus making it available for agricultural or municipal use. Where water is in short supply, the dual purpose use of hot water resources for power and for water has attracted serious consideration. Because water from the wells is already heated, desalting by distillation techniques similar to those being developed for seawater may prove relatively inexpensive.

Alternatively, the hot water itself could be used to generate power by systems in which the heat is transferred to a secondary fluid, which is then passed through power turbines. Use of secondary power fluids like isobutane which have lower boiling points than water is of particular interest for resources with temperatures less than about 200°C, where ordinary steam turbines become ineffective.

Power turbines designed for isobutane are typically smaller and have fewer stages than steam turbines—a consequence of the higher density of isobutane vapor. Because groundwater does not come in contact with the turbine, corrosion-resistant materials are not needed in their construction. Because isobutane is flammable, however, the secondary fluid system must have very tight seals and other safety features unnecessary with steam. Some heat is lost during its transfer from the hot water to isobutane, but less than is lost by extracting steam from the water, according to J. H. Anderson of York, Pennsylvania, a consultant to Magma Energy, Inc. Prototype turbines for use in geothermal power plants have been developed by Magma Energy for San Diego Gas and Electric, which has plans to build a 10-Mw prototype unit near the Salton Sea. Costs for power plants based on secondary fluid technology are uncertain and vary with the temperature of the water supply and the cooling methods used, but in favorable conditions they are estimated by Magma Energy to be competitive with fossil-fueled plants. A geothermal power plant in which secondary fluid turbines are used is now operating in the U.S.S.R.

Proven sources of hot water are located in California, Nevada, New Mexico, Oregon, and Idaho, although none is yet in commercial use, and potential sites have been identified in all of the western states. In Mexico, just south of the U.S. border near Cerro Prieto, a 75-Mw power plant that uses steam separated from hot water is about to begin operation. The Imperial Valley in California has been intensively investigated by Robert Rex, of the University of California at Riverside, and others, and this region appears most likely to be developed rapidly. Rex has located some eight thermal anomalies in the valley and has estimated the potential resource to be as large as 20,000 Mw of generating capacity and over a billion acre-feet of water, much of it in the southern portion of the valley where the water is expected to be less brackish than that near the Salton Sea. The Bureau of Reclamation of the Department of the Interior has drilled a test well at one of the hot areas in the region and plans further exploration.

Underground water systems do not come in contact with most of the near-surface deposits of geothermal heat. Hot rock is therefore a much larger resource, but one that is more difficult to exploit than steam or hot water. One proposed method of tapping these "dry" geothermal deposits is to create artificial cavities by means of conventional or nuclear explosives, and then to circulate water from the surface through the cavities to extract heat from the rock. Uncertainties about the seismic effects of the blast waves and about the economic feasibility of extracting heat from the small cavities that result have so far prevented attempts to develop the method in any detail.

Another method of getting at dry geothermal deposits, which has attracted considerable attention, originated with a group headed by Morton Smith at the Atomic Energy Commission's Los Alamos Scientific Laboratory (LASL) in New Mexico. They propose to use hydrofracturing techniques similar to those employed in petroleum recovery to create large cracks in a bed of a hard rock such as granite. The cracks would expose a large area of rock to a circulating flow of pressurized water pumped down one well and up another to extract the heat. At the top of the well, the heat would be transferred to a secondary fluid before being delivered to a turbine. Field trials are planned at a site on the edge of an extinct volcano near Los Alamos.

The LASL method, if it were successful, would make available geothermal resources estimated to be at least ten times the total from steam and hot water. There are uncertainties, however, about how well the hydrofracturing techniques that were developed for cracking sedimentary formations will work in the harder igneous rocks. Other unknowns are whether the granite will be impermeable enough to keep pressurized water from leaking away and whether it will contract as it cools in such a way as to extend the initial cracks, making new hot rock accessible and perpetuating the useful life of the system. How granite behaves at the temperatures and pressures in question is not very well known and there is no agreement among earth scientists as to the likelihood of the method's success. Preliminary calculations, however, make the LASL group confident that hydrofracturing or some similar technique will work.

Whether dry geothermal resources can be economically competitive with other sources of energy will depend in part on how deep the deposits are, since the expense of drilling through hard rock is expected to be a major cost. But deep drilling may not be necessary to develop sizable resources. The LASL team expects to reach temperatures of about 300°C at depths of 2300 meters at their test site, which is located on the boundary of a magma chamber. David Blackwell of Southern Methodist University recently discovered an area near Helena, Montana, that shows even greater promise as a shallow heat deposit. From measurements of heat flow in mines and wells, he estimates a heat source with a radius of about 4 kilometers and with temperatures between 500 and 700°C, the top of which is 1 to 2 km below the surface. Blackwell found no surface manifestations of the heat source, which he believes to be a recent magma chamber that never erupted to the surface; he points out that much more extensive heat flow measurements will be required to systematically explore for shallow deposits. Improvements in drilling technology could make even deep deposits of heat economical to tap.

The most likely course of development for geothermal energy, most observers believe, will be the rapid construction of power plants using steam separated from

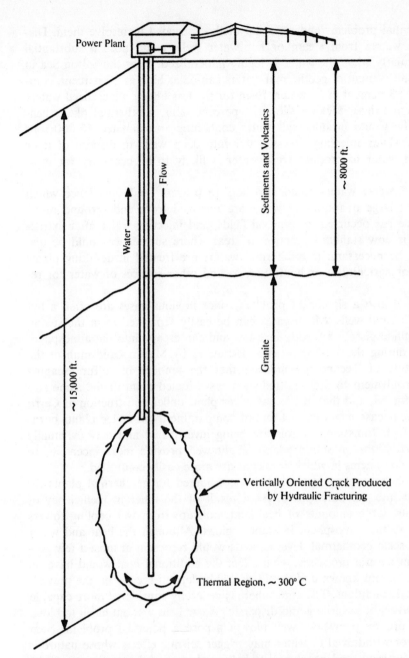

Power Plant

Flow

Water

~ 15,000 ft.

Sediments and Volcanics

~ 8000 ft.

Granite

Vertically Oriented Crack Produced by Hydraulic Fracturing

Thermal Region, ~ 300° C

FIGURE 19. Proposed system for extracting energy from a dry geothermal reservoir.
[*Source: Morton C. Smith, Los Alamos Scientific Laboratory.*]

high-temperature water deposits, with subsequent exploitation of low-temperature water reservoirs and dry geothermal deposits. Use of secondary fluids like isobutane rather than steam to drive power turbines is expected to be necessary for low-temperature water and hot rock systems, and may be desirable even at temperatures above 200°C to prevent corrosion and the release of environmentally undesirable substances.

Environmental effects of geothermal energy use have received considerable attention. Both environmentalists and advocates of geothermal power appear to recognize the potential problems and agree on methods needed to resolve them. Disposal of waste waters from steam or hot water wells could pose a substantial problem, particularly when the water is highly mineralized. Near the Salton Sea in California the salt content of geothermal waters can be as high as 20 percent, compared to about 3.3 percent in seawater. Even for the less highly mineralized waters found near Cerro Prieto, Mexico (about 2 percent salt), geothermal plants generating 1000 Mw would produce salt water containing an estimated 12,000 tons of salt per day. Thus injecting excess water into deep wells to dispose of it or treating the salt water to produce fresh water is likely to be necessary for most U.S. facilities.

Reinjection of waste waters could also help to prevent land subsidence which may result when large quantities of water are removed from underground reservoirs. Subsidence has occurred in some oil fields, and injecting water after extraction of the oil is now standard practice in areas where subsidence could be disastrous. It may be necessary, if geothermal waters are treated in desalting plants and delivered for agricultural or home use, to find other sources of water for reinjection.

Air pollution is also a significant problem, since noxious gases are often a by-product of geothermal wells. Many gases can be easily separated from the steam, but hydrogen sulfide dissolves readily in water and can escape into the atmosphere by evaporation during the cooling process. Estimates by Martin Goldsmith of the California Institute of Technology indicate that the amount of sulfur escaping the Geysers is equivalent to that emitted by a fossil-fueled plant of the same size burning low-sulfur oil, and that at the hot water plant under construction at Cerro Prieto, the sulfur release might exceed that of comparable fossil-fueled plants burning high-sulfur fuel. Emission controls are being investigated and may eventually be required; restrictions on sulfur release might well provide more incentives to use secondary fluid systems in which emissions are more easily controlled.

Virtually all of the cooling methods being considered for geothermal plants release waste heat into the atmosphere. As a result of the inherent inefficiency of geothermal plants, large amounts of heat and moisture from wet cooling towers could be added to the atmosphere in some regions. Although the heat and water from even large-scale geothermal developments would represent at most a few percent of that from natural processes, what effect the additional heat would have on the local weather is not known and would probably depend greatly on the prevailing meteorological conditions. The atmosphere is generally considered more capable than lakes and rivers of absorbing and dispersing waste heat without harm to biota.

As with any drilling operation, well blowouts pose a potential problem. Near faults, injection or withdrawal of water may trigger seismic effects whose nature is not yet well understood, and geophysicists believe that careful monitoring will be necessary. Neither of these problems is considered a serious obstacle to geothermal development.

Geothermal power is not likely to replace either fossil fuels or nuclear fission as major sources of electricity, at least in the near future. But conservative estimates are that 100,000 Mw of generating capacity, a not inconsiderable resource, could with vigorous efforts be developed by the end of this century. The additional prospect of furnishing desalted water for hard-pressed arid regions is also attractive. The remaining problems require concerted effort as well as substantial sums of money for exploration and technology development. The prospects, however, seem well worth the price.

10.
SOLAR ENERGY
The Largest Resource

Not long ago, proposals for using the sun's energy were apt to be received with considerable skepticism. Within a few agencies of the federal government and at an increasing number of university and industrial laboratories, that is no longer the case. Indeed, perhaps the most impressive testimony to the prospects for this type of energy is the score of prestigious scientists and engineers who have begun working on methods for converting the sun's radiation into forms more useful to man—heat, electricity, or chemical fuels.

Within 5 years, many of these scientists believe, solar-powered systems for heating and cooling homes could be commercially available at prices competitive with gas or oil furnaces and electric air conditioners. More significant, but farther in the future, is using heat from the sun to generate electricity; experimental solar-thermal generators have been constructed in several countries, and several groups in the United States are designing systems that take advantage of improved materials and manufacturing techniques. Eventually the direct conversion of solar radiation to electricity by means of photovoltaic cells or its bioconversion to wood, methane, or other fuels on a large scale may become economically feasible.

Solar radiation is the most abundant form of energy available to man, and is so plentiful that the energy arriving on 0.5 percent of the land area of the United States is more than projected total energy needs of the country in the year 2000. Sunlight is diffuse and intermittent, however; to harness it for power will require large areas to collect sufficient amounts of energy and, for most applications, the means to store energy. Despite its abundance, solar energy has not been exploited except in a limited way in water heaters, furnaces, and space applications; the technologies that would allow more widespread use are not commercially available. Systems for heating and cooling houses or for generating electricity with sunlight could be built now, but they would cost more than comparable systems that burn fossil fuels. For some applications, however, disparity in cost could rapidly disappear as solar technology improves and as the costs of fossil fuels rise.

Whether or not solar energy becomes generally available in the near future, there is growing agreement that it will be important in the long run. That being the case, proponents believe that it is the most underfunded area of research in the energy field, accounting for less than 1 percent of federal research expenditures on energy.

Of the proposed uses of solar energy, heating and cooling for homes and low-rise commercial buildings are the most developed and will almost certainly constitute the first significant use of solar energy in this country. Solar water heaters are already in commercial use in Florida and in several countries overseas. Experimental houses have been equipped with solar heating systems and preliminary development of cooling systems has begun.

For space heating, the solar collector is typically a black metal surface that readily absorbs sunlight and is covered with one to three panes of glass to reduce the heat loss. The glass transmits incoming sunlight, but absorbs the longer wavelength radiation emitted by the hot metal, so that a "greenhouse" effect is created and the effectiveness of the collector is increased. The heat is held by water or air that circulates through the collector during the day, and part of it is stored for release at night or in bad weather. Hot water, hot rock, and chemical (change of phase) storage systems have been experimentally tested.

For air conditioning, refrigeration systems that depend on absorption of the coolant fluid appear to offer the best choice. Experimental cooling units are being developed by several university and industrial research groups. In prototype systems such as that developed by Erich Farber at the University of Florida, heat from the sun is used to drive ammonia from an ammonia-water solution, and the ammonia is collected and condensed. When cooling is needed, the liquid ammonia is allowed to evaporate and expand as in a conventional cooling system, and the spent vapor is reabsorbed in water.

For absorption refrigerating systems to work smoothly, temperatures around 120°C or higher are needed, and thus solar collectors that are more efficient than those for heating purposes alone will be required. One possibility may be surface coatings of the type developed in recent years for space applications, which emit very little of the solar radiation they absorb and which consequently attain higher temperatures than uncoated metal collectors. If such coatings can be produced on a large scale, their use might help to reduce the cost of solar heating and cooling, since collectors are the most expensive item of a solar energy system. Combined cooling and heating systems, which have not yet been built, are also expected to improve the economic prospects for both because of the joint use of the collector.

Substantial technical problems remain to be solved in the design of cooling systems, in the manufacture of surface coatings for collectors, and in combining solar heating and cooling systems. In most regions of the country backup systems based on conventional fuels will be needed for extended periods of bad weather. Nonetheless, one estimate indicates that if systems were commercially available now, solar heating would be cheaper than electric heating in nearly all of the United States and would be competitive with gas and oil heating when these fuels double in cost (Fig. 20). Proponents believe that solar heating and cooling systems could ultimately supply as much as half of the nearly 20 percent of total U.S. energy consumption that is now used for residential and commercial space conditioning and could reduce the peak use of electricity in summer.

Despite its advantages, acceptance of solar power may be slow unless what are essentially social problems can also be solved. As Jerry Weingart of the California Institute of Technology put it, "developing the technology is not enough," because the fragmented building industry is traditionally slow to adopt new techniques. Solar heating systems, despite their lower fuel costs, will entail higher initial costs, thus discouraging consumer acceptance. Some observers have suggested that governmental encouragement in the form of tax incentives or energy performance construction codes should be part of a national energy policy. The slow rate of replacement of housing, in any case, guarantees that several decades will pass before a new heating system could have a significant impact on total

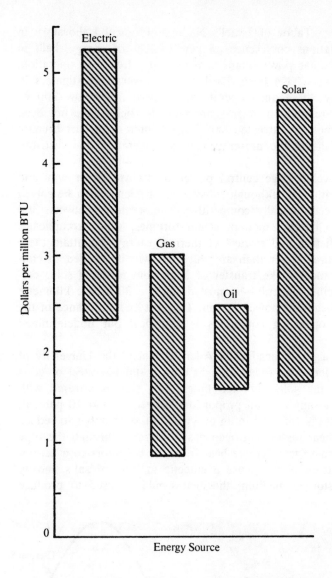

FIGURE 20. Costs of space heating with different energy sources.
[*Source: From* Solar Energy as a National Resource, *Department of Mechanical Engineering, University of Maryland, 1973.*]

energy use. Given the growing shortage of fossil fuels, however, it seems clearly advantageous to move in that direction.

Generating electricity with heat from solar energy is a more difficult challenge, and there are conflicting ideas about the best approach to the problem. Some engineers believe that small generating units located where the electricity is to be consumed are the ideal way to utilize a resource that is inherently diffuse and well distributed. This group favors the use of power turbines that would operate at temperatures considerably lower than those common in nuclear or fossil-fuel power plants, despite the low thermal efficiency, between 10 and 15 percent, that these units would have. Others have proposed large centralized solar-thermal facilities. The two concepts differ philosophically and technically.

Small vapor turbines that use heat from solar collectors to generate electricity

63

were demonstrated by Harry Tabor of Israel's National Physical Laboratory in Jerusalem at the United Nations conference on new sources of energy, held in Italy in 1961. A miniature solar power plant in Senegal is already in operation, and experimental solar engines have been developed by several investigators in the United States. Typically, these units operate at temperatures below 200°C. Their economic advantages relative to other sources of electricity have not been demonstrated, and the concept has attracted only limited interest, in part because of the difficulty of decentralizing the present electrical generation and distribution system.

Preliminary efforts to develop large central power plants are under way and have attracted considerable interest, although substantial problems must be solved before such plants can be economically competitive. Temperatures between 300 and 600°C are required to operate modern steam turbines, which complicates both collection of solar radiation and storage of thermal energy. To attain these temperatures, mirrors or lenses larger than any yet built will be needed to concentrate sunlight. Long distances make transfer of heat from far-flung solar collectors to the generating facility difficult—in most designs, a 30-square kilometer area is needed for a 1000-megawatt power station. The cost and endurance of the collecting apparatus under operating conditions is a critical but undetermined factor.

One design proposed by a group headed by Aden Meinel of the University of Arizona would use Fresnel lenses to focus sunlight onto a stainless steel or glass ceramic pipe concentrating the solar flux ten times. The pipe is covered with a selective coating that emits only a small proportion, between 5 and 10 percent, of the energy they absorb and is enclosed in an evacuated glass chamber to reduce conductive and convective heat losses. Nitrogen gas is pumped through the pipe at about 4 meters per second to transfer the heat from the collectors to a central storage unit. The Arizona team plans to use a eutectic mixture of salts, mostly sodium nitrate, as a heat storage medium; the heat would be used to produce

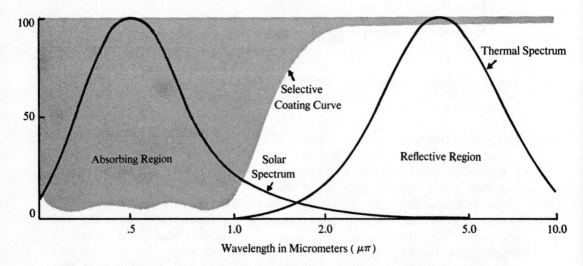

FIGURE 21. Spectra of incoming solar radiation and thermal reradiation from a heated surface. A selective coating captures the energy in sunlight by having a high absorptivity to visible light but a low emissivity (high reflectivity) to infrared light.
[*Source: Aden B. Meinel, University of Arizona.*]

steam for a turbine as needed. Liquid metal or the molten salt mixture itself, despite the greater difficulty in handling these substances, might be used in place of nitrogen to transport heat from the collectors to the storage unit.

A second group, headed by Ernst Eckert of the University of Minnesota and Roger Schmidt of Minneapolis-Honeywell, Inc., has also begun work on the central power station concept. Their design includes a self-contained, decentralized system for collecting and storing solar heat. A parabolic reflector would concentrate sunlight onto a heat pipe, a device that can transport heat along its length efficiently by convective processes and does not require a fluid to be pumped through it. The pipe's outer surface would be a selective coating, and the pipe would be enclosed in an evacuated chamber. A small heat storage tank attached to each heat pipe and reflector would complete the unit; no centralized heat storage facility would be used. Underground pipes would bring water to each storage tank and return it as steam directly to a turbine—thus reducing the pumping costs, the Minnesota team claims, compared to the nitrogen system. In addition, they believe, the self-contained system would be easier to construct and maintain.

The effectiveness of the collector's selective coating largely determines the temperatures that can be achieved. Two types of selective surfaces are known, both of which absorb much of the incoming radiation—in the visible region of the spectrum—but which emit only a small portion of the infrared heat radiation. Surfaces such as one developed by Minneapolis-Honeywell for the Air Force rely on optical interference between two reflective layers separated by a transparent layer. Thin films of this type have been routinely produced by vacuum coating techniques in the manufacture of tinted glass for the exteriors of new office buildings. A second type of surface, developed by B. Seraphin of Meinel's group at Arizona, is composed of silicon or similar materials that have natural selective properties. Layers of silicon and nonreflecting materials are laid down on a highly reflective substrate by chemical vapor deposition techniques; the silicon absorbs sunlight, but transmits infrared radiation, so that the composite surface has a high reflectance—and hence a low emittance—in the infrared.

These selective coatings are particularly important for planar collectors that are built without mirrors or lenses. Simple planar collectors have several advantages over the concentrating systems in that the concentrating collector must focus sunlight on the absorber and hence must follow the sun's motion in the sky; machinery to allow daily tracking complicates the collector design. In addition, focusing collectors operate only on direct sunlight, whereas planar collectors can utilize diffuse sunlight as well—and thus can function in cloudy or hazy weather. Because the performance of some of the most selective coatings decreases markedly at high temperatures, however, power plants using them would have to operate at temperatures below 350°C, with correspondingly reduced efficiency in the steam turbines. Improved selective coatings may allow planar collectors to be used. Meinel and his co-workers believe planar collectors to be in principle the most effective in areas of the United States other than the cloudless Southwest. But most initial designs are based on the assumption that concentration of the sunlight will be necessary, and in these systems the fabrication, cost, and durability of the concentrators are the major concern.

Type of collector is not the only feature of the design of solar thermal plants still being debated. Even with concentrating collectors, it may be advantageous to operate the system at a reduced temperature, according to the Minnesota team. Their analysis shows increasing efficiency of the collectors, but decreasing efficiency of the thermodynamic cycle of the turbines as the operating temperatures are reduced, with the optimum temperature dependent on detailed design of the system and on the heat storage medium chosen. Heat pipes of the size they envision have never been built, and other hardware details remain to be considered.

Both groups of investigators believe that the cost of solar-thermal plants will be not more than two or three times what fossil-fueled or nuclear-generating plants cost now, and that rising fuel costs will eventually tip the balance in favor of solar-thermal plants whose fuel is essentially free. Before accurate estimates of costs can be made, they agree, more detailed engineering studies and some additional research are necessary. But Meinel, at least, believes that full-scale solar-thermal power plants could be built as early as 1985 with an adequate research effort. Other estimates are somewhat less optimistic, but a group of western utility companies is considering the development of a small solar-powered facility that could serve as a prototype for peak load applications.

Potential environmental problems are probably fewer for solar energy than for other major sources of energy, but they do exist. Collecting surfaces absorb more sunlight than the earth does, and while small-scale use is not likely to alter the local thermal balance, the larger expanse of collecting surface in a central power plant might. Thermal pollution will also be a problem if water-cooled turbines are used—indeed, more so than with nuclear power plants because solar installations are expected to have even lower thermal efficiencies. If waste heat is returned to the atmosphere, it could help to restore the local thermal balance. The effects of small changes in the thermal balance would depend on the local meteorological conditions, but are expected to be small. Because of the lack of particulate emissions and radiation hazards, solar-thermal power plants might be built close enough to towns or industrial sites so that their waste heat could be put to use. Finally, like other industrial facilities, large-scale plants carry some risk of accident, with the attendant possibility of leaking corrosive heat transfer or storage media into the environment.

Space heating and cooling with solar energy are not available today. Solar-thermal power plants have yet to be built on any but the smallest scale, and key elements of the necessary technology have not been adequately demonstrated. But both options appear to be close enough to practical tests of their economic feasibility to warrant increased efforts. The ancient dream of power from the sun may be possible after all.

11.
PHOTOVOLTAIC CELLS
Direct Conversion of Solar Energy

Solar cells, which convert sunlight directly into electricity, are the predominant source of power for space satellites. Because they have no intermediate thermodynamic cycle, and because sunlight is a large and inexhaustible resource, solar cells are an inherently attractive source of power. At present, they are not competitive with other means of generating electricity for terrestrial use, but their long-run potential has attracted increasing attention. Recent improvements in cell manufacturing have stimulated novel proposals for utilizing solar energy on a large scale. Among the applications being studied are the solar-powered house—which some proponents believe could be a reality within 10 years—centralized generating stations, and, as a more distant possibility, large orbiting power stations that would transmit energy back to earth.

Of the difficulties that stand in the way of terrestrial applications, the key problems are reducing the cost of solar cells more than a hundredfold, increasing their useful lifetimes, and developing methods for storing energy. The relatively small numbers of cells produced for spacecraft are manufactured by costly batch processes and assembled by hand. Radiation damage in space or corrosion by humidity and other environmental agents on earth typically degrades the performance of unprotected cells well before the 20 to 30 years expected of power plants, and there is general agreement that encapsulation of cells in glass or plastic will be necessary. Space power stations do not suffer from the intermittent availability of sunlight, but for them to become feasible, the cost of transporting the components into orbit will have to be greatly reduced.

Despite these difficulties, there is considerable optimism among those working on photovoltaic cells, which have the advantage, they point out, of avoiding virtually all of the environmental contamination problems associated with other sources of power. New and unfamiliar technologies are not required—photovoltaic cells were among the first semiconductor devices to be developed. But in recent years, according to a report by the National Academy of Sciences, there has been very little support for research aimed at improving solar cells and also no financial incentive for major industrial development efforts. The report concludes that the efficiency with which existing silicon solar cells convert sunlight to electricity, about 13 percent, might well be increased to 20 percent.

Positive and negative charges are generated in solar cells by the absorption of photons. The charges diffuse across the cell until they either recombine or are separated and collected by an electrical inhomogeneity, typically a junction between two semiconductor regions. Silicon cells have been the mainstay of space power

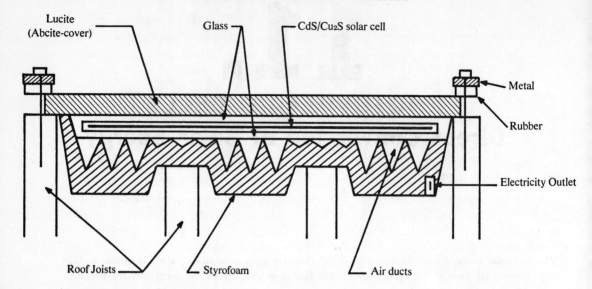

Lucite
(Abcite-cover)

Glass

CdS/Cu2S solar cell

Metal

Rubber

Electricity Outlet

Roof Joists

Styrofoam

Air ducts

FIGURE 22. Solar panel for deployment on roof joists, replacing normal roofing material. The panel collects both electrical energy (with solar cells) and heat energy (which is transported to storage systems by streams of air) for household use.
[*Source: Karl W. Boer, University of Delaware.*]

systems, but cadmium sulfide and gallium arsenide cells have also been developed and tested. Existing silicon cells develop about 0.5 volt, so large numbers of cells must be arranged in series to achieve high voltages. The output is direct current, which in a practical power system would have to be converted to alternating current before distribution to the consumer.

Silicon solar cells are made from single crystals of silicon, and production of these crystals has always been a costly and awkward process. In a promising development, however, continuous ribbons of silicon have been grown at Tyco Laboratories in Waltham, Massachusetts. In the Tyco process, which was originally developed for continuously growing single crystals of sapphire, the ribbon is formed from a die that controls the shape of the emerging crystal. According to A. I. Mlavsky of Tyco, the process can produce crystals at the rate of more than 2 centimeters per minute and could easily be automated to produce 100 crystals simultaneously. Although the process works well for sapphire and has produced silicon crystals with very few dislocations, the silicon is not of high enough quality for solar cells. The problem is that molten silicon is a highly reactive substance, very nearly a universal solvent, and it appears to be dissolving part of the die, thus introducing impurities. A search for better die materials is continuing, and, if successful, could lower manufacturing costs of silicon solar cells.

Raw material costs are also high. Although silicon is an abundant element and is available in metallurgical grade at $600 per ton, the cost of the extremely pure material needed for cell manufacture is 100 times more, and its production consumes large amounts of energy. Assembling solar cells into large arrays is also expensive and not easily automated. Although some have speculated that costs of energy from silicon solar cell assemblies may rapidly be reduced to less than the $400 per kilowatt typical of nuclear power plants, others believe that additional breakthroughs in manufacturing and materials will be necessary to reach that goal.

The cadmium sulfide cell appears to be a more likely candidate for low-cost photovoltaic cells within the near future, although it has the disadvantage of considerably lower efficiency. It converts only about 6 percent of the incident light to electricity. The main advantage of the cadmium sulfide cell is that it can be made from microcrystalline thin films, rather than from single crystals. These cells can be produced by vacuum-depositing thin layers on plastic, a process that lends itself to continuous mass-production methods. Several estimates suggest that cells cheap enough to be commercially competitive with other sources of electricity could be manufactured with existing techniques. A major problem with cadmium sulfide cells, however, is their lack of reliability—they degrade easily in the presence of moisture, at elevated temperatures, and possibly under the influence of light.

Gallium arsenide cells have also been proposed as a substitute for silicon cells, in part because of their resistance to radiation damage. Small gallium arsenide cells with efficiencies claimed to be as high as 18 percent have been tested at one laboratory. There is some question, however, whether gallium can be found in sufficient quantities to permit large-scale use of such cells, and it seems likely that they will be more costly than either silicon or cadmium sulfide cells.

Even if low-cost, high-efficiency solar cells could be produced, an effective energy system would not automatically come to pass. A particularly difficult problem is that of energy storage. Electrochemical storage is one possibility, but batteries of adequate capacity that can also withstand frequent charging and discharging for many years have yet to be developed. Hydrostorage—in which water is pumped uphill when power is abundant and allowed to flow through hydrogenerators at times of peak demand—is used by some utility systems, but is limited to a few regions of the country. Mechanical storage in high-speed flywheels is considered by some observers to be a realistic possibility, although little work on practical systems has been done. Another attractive proposal is to store energy in the form of hydrogen, which could be reconverted to electricity in fuel cells; the low-voltage, direct-current output of solar cells is ideal for electrolysis of water (from whence the hydrogen) but potential safety problems need to be resolved, as well as difficulties of storing gaseous hydrogen and of finding long-lasting and inexpensive catalysts for the fuel cells.

Because practical large-scale energy storage is not yet available, most proposals for photovoltaic power systems have been designed with limited storage. A prototype solar house being developed by Karl Boer and his colleagues at the University of Delaware in Newark, for example, will be connected with the existing utility system in a tandem arrangement. Sunlight would be a supplemental source of energy, providing electricity and heat to the house and to a small conventional storage battery during daylight hours and in essence supplying excess power to the utility network. In emergencies and during the hours of peak demand, the utility company, by switching the house to the storage battery, could lighten its load. Conventional power plants would provide the bulk of the electricity for the utility system. Because peak hours of sunlight coincide to some extent with the hours of maximum use of electricity, Boer believes that as much as 20 percent of a utility's power could be supplied by solar energy without the use of major storage facilities—in effect, solar cells on houses and commercial buildings would provide reserve capacity for the utility.

The solar cells in the Delaware design would be cadmium sulfide cells encapsulated in plastic panels that would replace normal roofing materials. Because of the low efficiency of the cadmium sulfide cells, virtually the entire roof of a house would be needed to provide adequate power, Boer estimates. In addition to generating electricity, the roof panels absorb heat that is collected and stored in

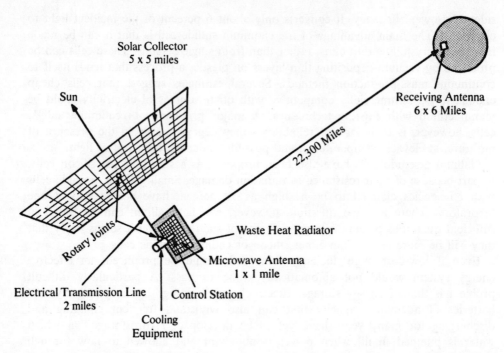

Figure 23. Proposed satellite solar power station designed to produce 10⁷ kw of electricity.
[*Source: Peter E. Glaser, Arthur D. Little, Inc.*]

compact thermal reservoirs of frozen salts. An electrically driven heat pump connected to one of the reservoirs heats and cools the house, and other reservoirs provide hot water.

Although the cost of the prototype solar system will be considerably more than for conventional systems, Boer believes that mass production of the solar panels, with only slight improvements of existing techniques, could bring the cost of power from the system to within range of commercial feasibility. Utilities, he believes, might be willing to partially subsidize the extra cost of a solar house and to maintain the solar panels in return for use of the excess power from the unit. Although costs and the technical feasibility of an interconnected system on a large scale remain to be demonstrated, the advantages of a distributed system and its potential for rapid development have attracted considerable notice.

A more long-range system that would also avoid the need for major storage of power is a space power station in synchronous orbit around the earth. As proposed by Peter Glaser of Arthur D. Little, Inc. in Cambridge, Massachusetts, large steerable arrays of silicon cells would generate electricity that would be converted to microwave power, transmitted to antennae on earth, and converted back to electricity. Large antennae, about 1 kilometer in diameter in space and at least 7 km in diameter on earth, would be needed to transmit and receive the microwave beam efficiently. Although such a system could provide large amounts of power, questions about the endurance of the components, the control of large structures in space, and the safety of the microwave radiation have still to be answered. Demonstration of such a system will be difficult and costly, since it is dependent on the existence of an inexpensive space shuttle.

Consideration of photovoltaic cells for terrestrial power systems has only just begun, and more proposals for practical systems can be expected. It seems likely that efficiencies of silicon solar cells can be improved to around 20 percent, comparable to the efficiencies expected from large solar-thermal power plants, and that mass production techniques will eventually lower their cost significantly; low-cost cadmium sulfide cells may be available even sooner. Although large amounts of photovoltaic power are not likely to be available in the near future, and although the development of this technology will undoubtedly require major federal funding, the potential resource is large. There appears to be no basic reason why direct conversion cannot become a reality.

12.

FUEL FROM WASTES
A Renewable Energy Source

At the same time that the United States has begun to consume far greater quantities of fossil fuels than can be produced domestically, it has also begun to produce far greater quantities of solid organic wastes than can be consumed by landfills and other conventional methods of disposal. A great amount of experimental work indicates that these wastes can be converted into synthetic fuels, thereby alleviating the waste disposal problem and providing a renewable energy source.

Many arguments in favor of this approach have been highly simplistic, however, and have ignored the difficulties of marketing low-value energy resources. Many proponents of conversion, furthermore, have frequently both overestimated the amounts of suitable waste materials available and underestimated the difficulties of collecting sufficient quantities to make operation of a conversion plant economically acceptable. While conversion of organic wastes to fuels is clearly an ideal way to dispose of the wastes, its value as an energy source will probably depend on improved efficiency of waste collection and supplementation of the wastes with specially-grown plant material.

The United States generates fantastic quantities of solid wastes—about 1.1 billion tons of inorganic mineral wastes and more than 2 billion tons of organic wastes each year. As recently as a year ago, waste conversion proponents argued that treatment of this organic waste could produce nearly 2.5 billion barrels of oil per year, roughly half this country's annual consumption. Newer estimates suggest that this number is somewhat exaggerated.

A study prepared for the U.S. Department of the Interior's Bureau of Mines (BuMines) by Larry L. Anderson of the University of Utah indicates that more than half the total weight of these wastes is actually water. In 1971, Anderson says, the total amount of dry, ash-free organic waste produced in this country was only 880 million tons (Table 6). More than 80 percent of the total, furthermore, was so widely dispersed that it could not be used; only about 136 million tons of dry organic wastes were readily collectable for conversion. This amount would have produced 170 million barrels of oil—roughly 3 percent of 1971 consumption of crude oil or 12 percent of imported crude. Alternatively, it could have produced 1.36 trillion standard cubic feet (scf) of methane, about 6 percent of 1971 consumption of natural gas. But development of successful techniques for using the wastes, he points out, will certainly accelerate their collection.

There are three major routes for conversion of these wastes to synthetic fuels: hydrogenation, pyrolysis, and bioconversion. Hydrogenation and pyrolysis have

Table 6. Amounts (in millions of tons) of dry, ash-free organic solid wastes produced in the United States in 1971.

Source	Wastes generated	Readily collectable
Manure	200	26.0
Urban refuse	129	71.0
Logging and wood manufacturing residues	55	5.0
Agricultural crops and food wastes	390	22.6
Industrial wastes	44	5.2
Municipal sewage solids	12	1.5
Miscellaneous	50	5.0
Total	880	136.3
Net oil potential (10^6 barrels)	1098	170
Net methane potential (10^9 cubic feet)	8.8	1.36

[Source: L. L. Anderson, Bureau of Mines]

been brought to the pilot and demonstration plant stages of development, respectively, and will probably be commercialized within this decade. Bioconversion has received only a minor research effort, however, and commercialization is unlikely before 1985.

The most highly publicized, and perhaps most promising, conversion method is the hydrogenation process developed by Herbert R. Appell and Irving Wender of BuMines' Pittsburgh Energy Research Center. This process might more aptly be termed a deoxygenation or chemical reduction process, since the principal reaction is abstraction of oxygen from cellulose—the primary component of organic wastes—by carbon monoxide and steam. Like pyrolysis and bioconversion, the process can be applied to virtually all organic wastes.

In the process, organic waste and as much as 5 percent of an alkaline catalyst such as sodium carbonate are placed in a reactor with carbon monoxide and steam at an initial pressure of 100 to 250 atmospheres and heated at 240 to 380°C for as long as an hour. Under optimum conditions, as much as 99 percent of the carbon content is converted to oil—about 2 barrels per ton of dry waste. In practice, more than 85 percent conversion is normally obtained, but because some of the oil must be used to provide heat and carbon monoxide for the reaction, the net yield is about 1.25 barrels per ton of dry waste.

The product is a heavy paraffinic oil with an oxygen content averaging about 10 percent and a nitrogen content that may reach 5 percent when manure is the starting material. Sulfur content is generally lower than 0.4 percent, well below the limits for heating oils applied in many cities. The energy value of the oil is about 15,000 Btu per pound. The widely used No. 6 fuel oil, for comparison, has a combined oxygen and nitrogen content of about 2 percent and an energy value of about 18,000 Btu per lb. The energy value of the raw waste varies from 3000 to 8000 Btu per lb.

The hydrogenation process has been thoroughly tested in the laboratory, and the Pittsburgh center is now operating a 480 lb-per-day continuous reactor to test the process on various potential substrates. Congress has appropriated $200,000 for expansion of the plant to handle 1 ton of animal wastes per day, and an additional $300,000 for design studies for a $1.75 million pilot plant to convert wood processing and logging wastes to oil. The expanded animal waste plant should be in operation by June of 1973, according to G. Alex Mills, chief of

Table 7. Typical properties of No. 6 fuel oil and oils made from urban wastes by the Garrett and BuMines processes.

Property	No. 6	Garrett	BuMines
Elemental analysis, weight percent:			
Carbon	85.7	57.5	77.0
Hydrogen	10.5	7.6	10.7
Nitrogen	}2.0	0.9	2.8
Oxygen		33.4	8.8
Sulfur	0.7–3.5	<0.3	<0.3
Energy value, Btu/lb	18,200	10,500	15,000
Specific gravity	0.98	1.30	1.03
Density, lb/gallon	8.18	10.85	8.58
Energy content, Btu/gallon	148,840	113,910	128,700
Pumping temperature, °F	115	160	140
Atomization temperature, °F	220	240	nd

nd = not determined.
[Source: Garrett Research and Development Company, Bureau of Mines]

Table 8. Recoverable materials in municipal solid wastes.

Raw material	Composition (weight %)	Commodity value ($ per ton)	Estimated recovery (%)	Potential value ($ per ton of waste)
Water	25			
Dirt and debris	2–4			
Inorganic products				
Magnetic metals	6–8	20–40	95	1.15–3.00
Nonmagnetic metals	1–2	100–200	95	0.95–3.80
Glass	6–10	12–20	80	0.60–1.60
SUBTOTAL				2.70–8.40
Organic products	50–60			
Pyrolytic oil		12–15	40	2.40–3.60
Pyrolytic char		4–40*	30	0.60–7.20
Pyrolytic gas		4–5	20	0.40–0.60
SUBTOTAL				3.40–11.40
TOTAL				6.10–19.80

*Potential value as activated carbon.   [Source: Garrett Research and Development Company]

Table 9. Comparison of methods for conversion of solid wastes to clean fuel.

Process requirements	BuMines	Pyrolysis	Anaerobic digestion
Form of feed	Aqueous slurry (15% solids)	Dried waste	Aqueous slurry (3–20% solids)
Temperature	320–350 °C	500–900 °C	20–50 °C
Pressure	100–300 atm	1 atm	1 atm
Agitation	Vigorous	None	Slight
Other	Uses carbon monoxide	None	None
Form of product	Oil	Oil and char	Gas
Yield (percent of original material)	23%	40% oil; 20% char	26% maximum
Heating value	15,000 Btu/lb	12,000 Btu/lb, oil; 9,000 Btu/lb, char*	23,800 Btu/lb
Percent of original heat content recovered in product**	65%	82% (60% if char not included)	77% maximum

*All of gas and ⅓ of char used to supply heat. ** Assumes heat content of dry waste is 8000 Btu/lb.
[Source: NSF/NASA Solar Energy Panel]

BuMines' division of coal, and the wood waste plant could be built by the end of 1974. A full-scale commercial plant incorporating the BuMines process for either animal or wood wastes could be in operation by 1980, he estimates.

Many problems remain to be investigated in the pilot plant operations. Foremost among them is evaluation of economic feasibility. BuMines has been extremely reluctant to calculate operating costs of a commercial plant by extrapolation from laboratory work, and has generally justified its work primarily on the value of the product oil ($4 to $5 per barrel) and the cost of conventional methods of waste disposal (as much as $10 per ton). Because of the need for pressurized reaction vessels, however, hydrogenation is clearly the most costly of the conversion technologies.

The principal technical problem to be resolved in the pilot plants is the improvement of methods for handling the solids and introducing them to the reactor under pressure. Other problems include refinement of the process for maximum oil yields, separation of oil from the solids, and minimization of pollution, including control of sulfur emissions and purification of process water. But balanced against any pollution from the conversion facility, of course, must be the tremendous decrease in other types of pollution resulting from safe disposal of the wastes.

The second major route for production of synthetic fuels is pyrolysis (destructive distillation). A major disadvantage of pyrolysis is that the process generally produces at least three different fuels—gas, oil, and char, for instance—thus multiplying recovery and marketing problems. Because pyrolysis is performed at atmospheric pressure, however, construction and operating costs should be lower than for hydrogenation. Several groups have investigated pyrolysis, including BuMines, but the most advanced process was developed by Garrett Research and Development Company, La Verne, California, the research arm of Occidental Petroleum Corporation.

The Garrett pyrolysis process is part of a complete system designed for disposal of urban refuse. Waste is first shredded and dried, and inorganic materials are removed for recycling or disposal. The organic waste is then reshredded and heated by a proprietary heat-exchanger to about 500°C in an oxygen-free atmosphere. Each ton of refuse produces almost 1 barrel of oil, 140 lbs of ferrous metals, 120 lbs of glass, 160 lbs of char, and varying amounts of low energy gas (400 to 500 Btu per scf). The gas is recycled to provide the oxygen-free atmosphere for pyrolysis and, with part of the char, is burned to supply heat for the process.

The oil thus produced contains about 33 percent oxygen, less than 1 percent nitrogen, and less than 0.3 percent sulfur. Because of the high oxygen content, the energy value of the oil is only about 10,500 Btu per lb. The specific gravity of the oil is about 1.3, however, compared to 0.98 for No. 6 fuel oil, so the energy content is actually about 75 percent that of fuel oil by volume.

Garrett claims to have invested more than $3 million in development of the pollution-free process and has tested it for more than a year in a 4 ton-per-day pilot plant at La Verne. The company recently received a contract to build a $4 million, 200 ton-per-day demonstration plant that will handle all the solid wastes produced by the communities of Escondido and San Marcos, California. The Environmental Protection Agency will provide 75 percent of the funds for the plant, which is scheduled to begin operation in November 1974. Oil from the plant will be sold to the San Diego Gas and Electric Company, which will invest $150,000 for new facilities to handle it. Operating costs for the plant are expected to be a little more than $8 per ton, about 20 percent less than conventional disposal costs in the area.

A full-scale, 2000 ton-per-day plant to process wastes from a city of 500,000 would cost about $12 million, Garrett estimates. In around-the-clock operation at a municipally financed plant, the firm contends, it would cost $5 per ton of refuse to produce oil and other recoverable products worth $6, even with no credit for waste disposal.

Several other groups—including Monsanto Enviro-Chem Systems Inc., St. Louis, Missouri; Torrax Systems Inc., Buffalo, New York; the Union Carbide Research Center, Tarrytown, New York; and Battelle Pacific Northwest Laboratories, Richland, Washington—are investigating pyrolysis of wastes to produce primarily char and low-Btu gas. With few exceptions, these processes have been developed principally for volume reduction of solid wastes, and production of gas has not been optimized. The low-Btu gas, furthermore, cannot be transported economically and must be used at the production site, which greatly reduces the viability of such processes.

The city of Baltimore, Maryland, for example, with EPA assistance is planning to install a $14 million, Monsanto-designed pyrolysis facility that will handle 1000 tons of solid waste daily. Fuel gas generated in the process will be burned on site to produce steam that will be sold to the Baltimore Gas and Electric Company for the production of electricity. Even so, operation of the plant is expected to cost nearly as much as would disposal of the wastes in a landfill. Furthermore, many conventional waste disposal incinerators throughout the country also generate steam, but have been unable to sell it. Oil production is thus a much more desirable alternative.

Production of methane that could be used interchangeably with natural gas is also a desirable alternative, but the cost of upgrading the low-Btu gas produced by pyrolysis is prohibitive. Methane can, however, be produced by the third major technology, bioconversion through digestion by anaerobic bacteria. Development of this process is less advanced than the others, but current work suggests that more than 10,000 scf of methane, with an energy content of about 1000 Btu per scf, could be produced from each ton of solid waste.

Anaerobic digestion has been used for many years to reduce and stabilize municipal sewage, and EPA is sponsoring some research for that purpose. Only recently has the National Science Foundation also begun to fund energy-directed digestion research, as part of its Research Applied to National Needs (RANN) program.

The principal recipient of such RANN funds is the University of Pennsylvania's Center for Energy Management, Philadelphia, which was awarded $600,000 for a 3-year study of the feasibility of converting urban wastes to methane. A probable participant in the project will be United Aircraft Corporate Research Laboratories, East Hartford, Connecticut, which has been investigating waste digestion for several years. According to Pennsylvania's Martin Wolf, the university group will study bacterial processes in the digester, control and maximization of gas production, fuel utilization, and environmental considerations, while United Aircraft will focus on systems design, component procurement, and operation. Other major investigators include John T. Pfeffer's group at the University of Illinois and Dynatech Corporation, Boston, Massachusetts.

Although bioconversion is theoretically a simpler process than hydrogenation or pyrolysis, a large number of problems remain to be solved. Among them are the need for new techniques to feed solids into the digesters and inexpensive ways to collect and purify the methane, recirculate the effluents, and control pollution. A major environmental problem is disposal of the organic sludge that remains after digestion, which may amount to 40 percent of the starting material.

This sludge could possibly be dried and burned, but that could produce air pollution. It could conceivably be converted to oil or gas with one of the other techniques, but it would then seem more logical to use that technique on all the waste. Or, since the sludge has a high protein content, it might prove valuable as a raw material for the manufacture of animal feed. Whatever the solution, though, the economics of sludge disposal will play a major role in determining the overall viability of bioconversion.

If the preliminary investigations suggest that these economics are favorable, a small NSF-funded pilot plant could be in operation within 5 years, according to the NSF's Lloyd Herwig. A larger, 10 to 100 ton-per-day demonstration plant, possibly funded jointly by federal and local governments and industry, could be in operation in 8 to 10 years. A full-scale commercial plant, he estimates, could be in operation in about 15 years.

As energy sources, however, all of the conversion technologies are severely restricted by the limited amount of solid waste available. A number of investigators have therefore suggested supplementing these wastes with algae, phytoplankton, and other plants grown specifically for conversion to fuels. Many have even suggested more direct use of such plants—fermentation of cereal grains to alcohol, for example, or direct combustion of wood from trees. For now, however, the costs of harvesting and transporting such plants appear far too great to justify such an approach, and in the future, the plants will probably be more valuable as food or lumber.

Even on the best land, moreover, the photosynthetic efficiency of farming—the amount of incident energy stored in the crop—is rarely above 0.5 percent. Much greater efficiencies can be obtained with fast-growing algae in ponds or the ocean. Between 4 and 10 percent of incident light energy can be stored by strains of algae that are now grown outdoors, but strains that are at least 20 percent efficient have been grown in the laboratory. Efficiencies can also be improved by enriching the algal medium with carbon dioxide and ammonia.

Most work on algal growth has been directed toward their use as an oxygen source in sewage treatment facilities, and even this research has been very limited. More data are necessary before it will be possible to determine whether algae can be grown cheaply enough for use as a raw material in fuel production. Major problems, many of which will be investigated by the University of Pennsylvania–United Aircraft project, include lowering the construction costs of facilities for growing algae; providing adequate, cheap nutrients, such as sewage, perhaps; and harvesting and transporting them to the conversion facility efficiently. Even by the most optimistic estimates, it will be at least 25 years before large-scale production of algae could begin to contribute at least 1 percent of this country's energy supply.

13.

MAGNETIC CONTAINMENT FUSION
What Are the Prospects?

Very early in the atomic age it was realized that the reaction that produces the hydrogen bomb could be a great source of energy if it could only be controlled. At one time it was thought that research on a fusion reactor might proceed so quickly that fusion would become an alternative to the first generation of breeder fission reactors. But the early projections were too optimistic. No one knew in the early 1950's how slow progress toward a fusion reactor would be because few scientists realized that first it would be necessary to unravel and master the details of a whole new field of science—plasma physics. Scores of different shapes for magnetic systems have been tested to see how well they contain the hot, ionized gas of a plasma for a fusion reaction. But so far none has shown that net production of energy is feasible.

Encouraging experiments in the last 4 years have led many scientists to think that controlled fusion is probably attainable with magnetic containment systems, possibly by 1980. Some scientists have estimated that another approach to fusion —with a laser to heat the fuel—might be feasible sooner. If the scientific feasibility of either magnetic or laser fusion were demonstrated, commercial sales of fusion reactors would still not begin until after experimental reactors were extensively tested and a demonstration reactor proved successful. The specific studies necessary to begin to assess the size, cost, operating characteristics, radiation hazards, and environmental effects of a fusion reactor are in a very early stage for laser fusion and are just becoming available for magnetic fusion. However, it is clear that fusion reactors would have two great advantages: virtually unlimited fuel resources and no conceivable danger of an explosive accident.

Two heavy isotopes of hydrogen, deuterium and tritium, are commonly considered as the likely fuels for fusion. Deuterium is so plentiful in seawater that it would be an extremely cheap fuel (costing only 0.003 mill per kilowatt hour); but tritium would have to be bred in a fusion reactor, much like plutonium can be bred in a fission reactor. As an ultimate resource, the energy available from fusion is surpassed only by the resource of sunlight, which, after all, is just fusion power from a far greater reservoir. Of the several possible fusion cycles, the deuterium-tritium fusion cycle is the only one even remotely limited by resources because the earth's lithium would have to be mined to manufacture tritium. But known world lithium resources are great enough to supply energy needs for 1 million years.

The temperature for fusing deuterium and tritium is so high that no material could contain the fuel without melting. But magnetic fields can act like bottles to keep the hot fuel from touching any walls. Three types of magnetic field designs

79

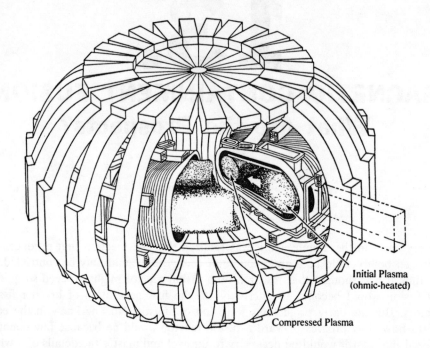

Initial Plasma
(ohmic-heated)

Compressed Plasma

FIGURE 24. Adiabatic Toroidal Compressor tokamak. It is one design for magnetically-containing a plasma to achieve a controlled fusion reaction. Compression of a plasma in this device has proved to be a viable method of heating fusion fuel.
[*Source: Princeton Plasma Laboratory.*]

seem to be quite promising: toroidal or doughnut shapes (the Soviet design, the tokamak, is the best-known), pinch devices, and magnetic mirror machines. There are reasons to think that increasing the size of each of the three devices would be sufficient to demonstrate scientific feasibility. But it won't be cheap. The total cost of U.S. fusion research necessary to build all three machines by 1980 would be about $1 billion, and some Atomic Energy Commission (AEC) officials have estimated that about $5 billion will be necessary to build an experimental reactor based on any one design. The budget for the Controlled Thermonuclear Research division of the AEC for fiscal year 1973 was $40 million.

The most recent designs for a fusion reactor have been based on a tokamak design for the magnetic containment system. In a tokamak reactor, a plasma gun would inject fuel every 3 minutes into a toroidal chamber where it would be heated and compressed into a dense plasma to produce fusion. In order to absorb the energy of the fusion reaction and to breed new tritium, the inner chamber would be surrounded with a blanket of lithium or lithium salts about 1 meter thick. Outside the lithium blanket a neutron-absorbing shield would be necessary to protect the coils of the superconducting magnets, which would be on the outside of the multilayer assembly. One meter of water would be an effective shield against neutrons, but if lithium came into contact with water a violent reaction would occur. Some engineers prefer a less hazardous material, such as graphite. The toroidal reactor would be a massive device, about 10 meters high and 30 meters in diameter.

Many design parameters for a fusion reactor are changing. Early design studies postulated an electrical power capacity of 5000 megawatts for a reactor based on a tokamak design, but the flux of neutrons passing through the inner wall proved to be so great that no material considered would last long enough to be practical.

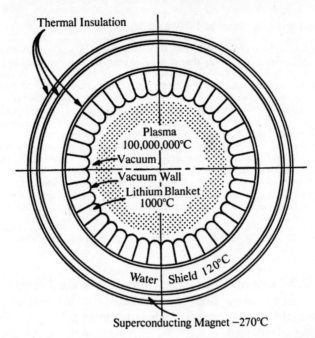

FIGURE 25. Cross-section of a proposed fusion reactor based on a tokamak device. Coils of the superconducting magnet would have to be very well-insulated from the extremely hot plasma in the center. A lithium blanket absorbs neutrons produced during fusion, converting their kinetic energy to heat. An additional layer of water is needed to protect the magnet from neutron damage.
[*Source: U.S. Atomic Energy Commission.*]

A recent design by Art Fraas, of the Oak Ridge National Laboratory, Oak Ridge, Tennessee, has a power of about 500 Mw, and a correspondingly lower neutron flux. Though the lower-powered reactors appear to have less troublesome materials problems, the capital costs and the operating costs will almost certainly be higher because the very expensive magnets will be used less efficiently. However, improvements in the technology of large superconducting magnets and efficiencies of mass production could reduce magnet costs according to Robert Hirsch, director of the Controlled Thermonuclear Research division of the AEC. All costs are quite uncertain now; estimates for the superconducting magnets alone range from $20 to $60 per kw of electrical capacity.

The design of a fusion reactor will present many difficult engineering problems because enormous differentials of temperature and neutron flux must be sustained over very small distances. Typically, the temperature at the center will be 100 million °C, and the neutron flux will be greater than 10^{13} neutrons per square centimeter per second. But only 2 meters away, where the superconducting magnets will be situated, the neutron flux and the absolute temperature must be almost zero.

One of the most troublesome engineering problems is the choice of material for the inner wall. It should not react with very hot liquid lithium (1000°C), and it should last at least 10 to 20 years in a very large flux of neutrons. (The neutron flux in reference designs for a fast breeder fission reactor cooled by liquid metal is several orders of magnitude greater, but the higher energy neutrons from a deuterium-tritium fusion reaction are much more damaging to the structure.)

Pure lithium would be desirable for a blanket because it breeds tritium more efficiently than lithium salts do, but it could be very troublesome to pump liquid

Table 10. Delineation of the Steps to Fusion Power.

(1) A *basic fusion plasma research and developmental phase* in which research experiments to produce, heat, contain, and study thermonuclear plasmas are conducted in parallel with the development of associated fusion technologies.

(2) A *scientific feasibility phase* in which experiments are constructed and operated which attempt to reach "break-even" fusion plasma conditions (minimum values of density, temperature, and plasma confinement time) in laboratory configurations which lend themselves to development into net power producing systems. Fusion fuels need not necessarily be used in these experiments. The AEC program is beginning to enter this phase.

(3) An *experimental reactor phase* in which one or more experimental reactors, designed to use fusion fuels and to produce net energy in a useful form (steam or electricity), would be constructed and operated.

(4) A *prototype* or *demonstration reactor phase* in which one or more electric power producing units, including all of the elements of a commercial power plant, would be built and operated. Successful operation of a demonstration plant would be a prelude to commercial sales.

[Source: Statement by Dr. Roy W. Gould to a subcommittee of the Joint Committee on Atomic Energy, Nov. 10, 1971.]

lithium through the system. Since liquid lithium is a conductor, it can only be pumped in certain directions without resistance from the magnetic field. Other ways to extract heat from the reactor, such as pumping helium through the lithium blanket, are being investigated.

The thermal pollution from a fusion reactor could possibly be less than from fossil fuel plants because the high operating temperature would make possible a large thermal conversion efficiency. The design proposed by Fraas would convert 55 to 60 percent of the thermal power to electrical power—compared to 40 percent for a modern fossil fuel plant—by use of a binary vapor generating cycle. However, efficiency could be reduced significantly if a very large amount of energy is needed to power a plasma gun for injection of the fuel.

Very high conversion efficiencies might be achieved by directly converting fusion energy into electrical energy, bypassing the thermal cycle. Electricity could be produced by allowing the charged particles in the plasma to move through an electric or magnetic field. However, the most likely fuel for fusion—a deuterium-tritium (D-T) mixture—would be the least practical for direct conversion because only 20 percent of the energy is released as charged particles. The feasibility of direct conversion for other fuels, such as deuterium plus deuterium (D-D) or deuterium plus helium-3, has not yet been shown, according to David J. Rose of the Massachusetts Institute of Technology, Cambridge, Massachusetts. For a mirror or pinch type of fusion reactor the direct conversion chamber would be a huge disk, about 100 meters in diameter, from which it would be very difficult to recover the spent plasma.

Operation of a fusion reactor would present several major hazards. The hazard of an accident to the magnet system would be considerable, because the total energy stored in the magnetic field would be 2×10^{11} joules, about the energy of an average lightning bolt. An even greater hazard would be a liquid lithium fire (6×10^{13} joules).

But the greatest hazard of a fusion reactor—whether the magnetic containment type or the laser type—would undoubtedly be the release of tritium, the volatile and radioactive fuel, into the environment. Tritium has a relatively short half-life, about 12 years, but it spreads rapidly, both because it is a light gas and because it can replace hydrogen in molecules such as water. The low-energy beta emission of tritium is relatively benign compared to the radioactivity from many fission products, however, and the biological hazard of tritium from a fusion plant would be much less dangerous than the hazard of iodine-131, the volatile product from a fission plant.

The problem with tritium is that it is very difficult to contain. The most optimistic estimates agree that at least 0.03 percent of the total inventory would probably escape from the reactor each year. A great deal more tritium would penetrate the structure, according to David Rose, but would normally not escape. Most metals that seem suitable for fusion reactor structures become very permeable to tritium at the working temperatures of a fusion reactor. In a preliminary appraisal of the tritium hazards, Fraas estimated that 60 curies of tritium would be released per day, one-fourth in the form of water released through the steam cycle. Tritium in water is much more dangerous than tritium gas because it is more rapidly assimilated by the human body. But if the entire cell in which the reactor is contained were evacuated, Fraas thinks that the total tritium release could be

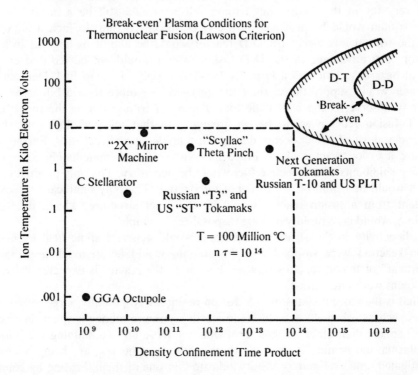

FIGURE 26. A graphical presentation of the criteria for a practical fusion reactor. For a feasible fusion reactor, two criteria must be met: the ion temperature of the plasma must be high enough to initiate fusion and the plasma must remain dense long enough to extract net energy. The figure illustrates how close various fusion devices come to meeting the criteria. For a deuterium-tritium (D-T) fuel, net energy or "break-even" can be achieved at a temperature of 100 million degrees and a product of density and confinement time of 10^{14}, as indicated by the dotted lines. This is called the Lawson criterion. For D-D fusion the criteria are more difficult to meet. The three points indicate the operating conditions of three different types of U.S. devices: the 2X mirror machine at Lawrence Livermore Laboratory, the Scyllac sector-shaped pinch machine at Los Alamos Scientific Laboratory, and the ST tokamak at Princeton Plasma Laboratory. Next generation tokamak devices, both U.S. and Soviet, are expected to approach within half an order of magnitude of "break-even."

[*Source: Adapted from material supplied by the U.S. Atomic Energy Commission.*]

held to 1 curie. He emphasizes, however, that any estimate of the tritium release is necessarily uncertain because many of the necessary data are not known.

One of the unanswered questions about fusion power is whether the tritium hazard would be small enough to permit fusion reactors in the middle of cities, where much electric power is consumed and where waste heat can be sold for industrial and home use, rather than discharged to create thermal pollution. If an exhaust stack were used for a fusion reactor, the maximum exposure anyone would receive at ground level would be one-fifth the current AEC tolerance, according to Robert Hirsch. Such estimates are encouraging, but the leak rates assumed are near the lowest that have been achieved in any technology. Because the total tritium inventory of a fusion reactor will be about 100 million curies, reactors probably will not be located in urban areas until extensive operating experience has been accumulated.

A fusion reactor that burns deuterium (D-D) might become feasible at some time very far in the future, but tritium release would still be a hazard because some tritium would be produced in the D-D fusion reaction. In fact, it is possible that the tritium inventory of a D-D reactor would be almost as large as in a D-T reactor. The advantage of the D-D fusion reactor would not be reduced environmental hazard so much as a virtually limitless supply of cheap fuel. Deuterium in seawater would supply the world's energy needs for more than a billion years.

Many observers believe that the total release of tritium during the fuel cycle of a D-T fusion reactor would be no greater than that released during the fission fuel cycle, most of which occurs during the reprocessing of fuel rods. Furthermore, because a fusion reactor would breed its own fuel—and immediately burn it—no massive shipments of radioactive fuel would be necessary. The only product of D-T fusion would be helium-4, which is not radioactive. The only significant radioactive remnant from a fusion reactor would be the inner structure, which, if made of niobium, would cool rapidly making disposal fairly simple.

Radioactivity in the structural members would generate some heat even if the fusion reaction were stopped, but the amount would be so much less than the "afterheat" of fission reactors that no damage to the reactor is expected if the flow of coolant were interrupted.

What is the current status of U.S. fusion research? Apparently, large toroidal machines such as tokamaks are closest to feasibility; more money is spent on research for tokamaks than other designs. Problems for these are maintaining the purity of the plasma and heating it. At least three methods of heating are being intensively investigated, and preliminary results indicate that one of them, heating by compression, is feasible. It is being investigated with the adiabatic toroidal compressor tokamak at the Princeton Plasma Laboratory.

Magnetic mirror machines will only work if a quiescent plasma state can be reached. Three approaches are being tried; a promising one, Baseball I at the Lawrence Livermore Laboratory, uses neutral beams for heating. If a quiescent or steady plasma state can be reached, and if the losses of plasma through the ends of the mirror agree with theory, the next step would be a machine large enough to demonstrate feasibility. Substantial upgrading of the neutral beams would be required, however.

Large pinch machines, such as the Scyllac machine at Los Alamos Scientific Laboratory, will be thoroughly tested about 1975. However, such machines need large amounts of power delivered very rapidly and the present power sources, banks of capacitors, will not be sufficient for a feasibility-testing machine. If the

Scyllac proves successful, different power sources—large inductors, for instance—and two-step compression and heating will be required.

No dramatic breakthroughs have occurred recently in fusion by magnetic confinement, and controlled fusion is still far from a certainty. However, many scientists think that scientific feasibility will be proved within the decade. If fusion works, some observers expect that the development of a commercial power plant will be rapid. Others, however, note that, in contrast to the history of fission power (less than 3 years elapsed between the idea of a fission chain reaction and the proof of scientific feasibility with a simple assembly of graphite braced in a wooden frame), almost 20 years have elapsed since the beginning of research to prove the feasibility of fusion. If the complexity of the feasibility experiments is any indication of the sophistication of a future fusion reactor, development of a commercial reactor may be as halting and tedious as has been progress toward providing the principle.

14.

LASER FUSION
A New Approach to Thermonuclear Power

If the fusion of hydrogen taking place in the sun could be controlled on earth, seawater could provide fuel for the world for a million years or more, but no device for controlling fusion has been proved feasible. For the last 20 years, most of the effort and money for U.S. fusion research has been directed toward the goal of containing a fusion reaction with magnetic forces, but many scientists now think that the goal of achieving fusion with high-powered lasers could be reached first.

After Soviet researchers demonstrated in 1968 that a laser could initiate a fusion reaction, many countries expanded their laser research programs. But the size of the lasers needed for a practical power plant was considered impossibly large until the U.S. Atomic Energy Commission (AEC) in 1971 revealed details of a new laser method that had previously been classified. Instead of heating a pellet of hydrogen isotopes with a single laser beam, as assumed in most older calculations, many beams from different directions will heat the pellet in such a way that it is greatly compressed. The advantage of the new scheme has so far been demonstrated only in computer simulations; but if the predictions are borne out in experiments, large lasers now under construction in the Soviet Union, or planned for construction within the next year or two in the United States, could prove the scientific feasibility of the principle.

Some scientists predict that the feasibility of laser fusion will be proved in two to four years. This would contrast dramatically with the slow progress of magnetically confined fusion. However, research on magnetically confined fusion is much more advanced, in the sense that there is a long history of comparison between theoretical predictions and experimental results. The big lasers necessary to test the feasibility of laser fusion are just now becoming available, and lasers for a practical power plant would require much more development.

The general outline of the laser fusion process is that a small, well-focused pulse of laser light very rapidly heats a small pellet of deuterium and tritium to temperatures hot enough to ignite fusion (100 million °C) before the pellet can expand significantly. Because Newton's law of inertia largely determines the speed of expansion (about 10^6 meters per second), laser fusion is often called inertially confined fusion. A 1-millimeter pellet expanding at 10^6 m/sec would double in size in 1 nanosecond (10^{-9} second), so the laser energy must be delivered very rapidly.

U.S. research on laser fusion is proceeding along two parallel paths, with little interaction between them so far. Theoretical efforts have consisted of computer simulations of the complicated interaction between the laser and fuel pellet and experimental efforts have concentrated on developing powerful lasers capable of delivering their energy in less than a nanosecond. Because neodymium glass lasers have focusing and timing characteristics superior to other types of high-powered lasers, they are receiving the most attention.

The Soviet Union has the largest research program for laser fusion; it is estimated to be twice the size of the U.S. program. But the U.S. program has recently been enlarged. Federal budget requests for laser fusion research rose from $13 million in fiscal year 1972 to $23 million in fiscal year 1973, and are expected to continue to increase. Intensive programs of research are being mounted at the Lawrence Livermore Laboratory and the Los Alamos Scientific Laboratory. Private sources are supporting laser-fusion research at the University of Rochester, where the university, General Electric Company, and Exxon Research and Engineering Company sponsor a program that cost about $1 million in 1972, and at Ann Arbor, Michigan, where KMS Industries, Inc. plans to spend $50 million in four years. Although the hope and enthusiasm of the private sector is for an early demonstration of fusion for power generation, the focus of the AEC research, administered under the Division of Military Applications, is not yet differentiated between military and civilian goals. One possible military application is a pure fusion thermonuclear weapon, which does not require uranium or plutonium.

The concept that has made many scientists more optimistic about the eventual success of laser fusion is an implosion in which the fuel pellet is compressed so that it burns much more completely. When several laser beams hit a pellet from many sides, the outer portion ionizes, heats up, and blows off at high velocity to produce large pressures in the center of the pellet. Using implosion could reduce the laser energy required for a practical power reactor by a factor of 1000, to a value of 10^5 or 10^6 joules, according to John Nuckolls and Lowell Wood at the Lawrence Livermore Laboratory, Livermore, California. However, this is still much more than the energy of the largest laser available today, a nine-path neodymium laser developed by Nikolai Basov and his co-workers at the Lebedev Institute in Moscow, which can deliver 600 joules in 2 nanoseconds.

In experiments to date the amount of fusion energy released has been much less than the laser energy, but Nuckolls and his colleagues estimate that a fusion reaction that exactly replenishes the energy of the laser light could occur at 10^3 joules. This balance of energy, called "break-even," is normally referred to as the criterion for judging the arrival of laser fusion at the point of scientific feasibility. It may be a deceptively easy milepost to reach, however, because typically only 0.1 percent of the electrical energy used by a neodymium glass laser emerges in the energy of the light emitted.

Achieving break-even will not herald the imminent arrival of practical laser fusion, but most observers agree that experiments at break-even power levels are exactly what is needed to check the difficult calculations of laser-pellet interactions, and to learn which of the many possible physical mechanisms will be most important. Details such as the amount of heating and compression are predicted from nonlinear calculations which have uncertainties whose cumulative effects may be quite large. Most diagnostic information about the laser-pellet interaction has been obtained by measuring the amount of reflected laser light and the number of neutrons produced in fusion, but both of these are relatively primitive measurements (some scientists think that the information about neutron yields is almost useless).

Apparently none of the U.S. laboratories has a laser configuration that could produce an implosion. French scientists at Limeil have a four-path laser that will produce 150 joules on a deuterium target, but most observers believe that only the

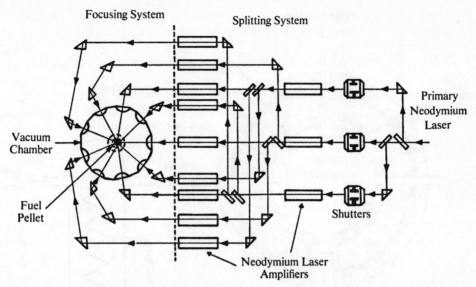

Focusing System **Splitting System**

Vacuum Chamber

Fuel Pellet

Primary Neodymium Laser

Shutters

Neodymium Laser Amplifiers

FIGURE 27. A schematic of a nine-path Soviet laser. A pellet of fusion fuel, which is placed where the beams converge at the center of the vacuum chamber, is very rapidly heated by the converging laser beams. The laser, which was constructed at the Lebedev Institute in Moscow, can deliver 600 joules of energy in a 2 nanosecond pulse. The primary beam is split into 3 parts, each of which is amplified by a larger laser consisting of a rod of neodymium-doped glass. Then each beam is again divided into three, and again amplified. A 27 beam laser, now being constructed by the Soviet group, is expected to yield 2000 joules in 2 nanoseconds.
[*Source: From* Physics Today, *August 1972.*]

group at the Lebedev Institute currently has the capability to begin testing the implosion theories. A 27-beam laser under construction there is supposed to produce about 10^4 joules and may be the first instrument to achieve a break-even reaction.

Research programs at U.S. laboratories are exploring both timing variations and wavelength variations. Researchers at Livermore and at KMS Industries, Inc. attempt to maximize the compression by time-tailoring the laser pulse very carefully. In some cases a small precursor pulse is used to create a thin plasma on the outside of the pellet. Certain effects in the thin plasma increase the symmetry of the implosion, so that high pressures can be reached.

Researchers at the Los Alamos Scientific Laboratory, the University of Rochester, Sandia Laboratories, and Livermore plan to experiment with wavelengths different from the near infrared radiation of a neodymium laser [1.06 micrometers (μm)]. Computer simulations of plasmas, performed at Livermore and Limeil, indicate that radiation at wavelengths shorter than 1 μm may be desirable for fusion. Shorter wavelength laser pulses are absorbed much deeper in the pellet, and thus are more effective in generating an implosion.

Although neodymium lasers are excellent sources of powerful short pulses, it is doubtful that they could ever be used to drive a power plant because of their poor efficiency. Neodymium lasers are also difficult to operate at high repetition rates, and scientists at both Livermore and Los Alamos estimate that fast repetition rates (one shot per second or faster) will be needed for a useful power plant. At the current state of technology, the durability of neodymium systems is also far too poor to be practical.

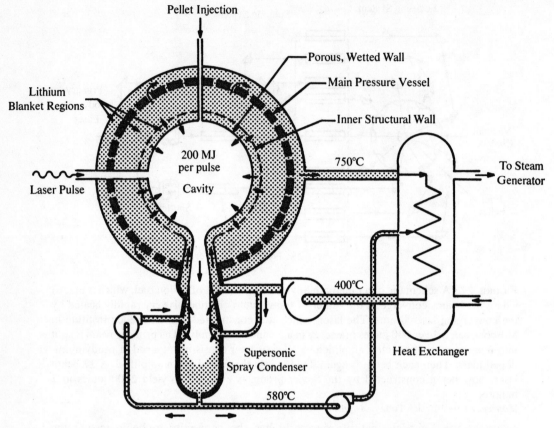

FIGURE 28. Laser fusion reactor with a "wetted-wall." This proposed design employs a porous wall to define a cavity at the center of a sphere of liquid lithium. Fuel pellets are injected into the cavity through a channel at the top and heated by a laser beam aimed through another channel at the side. A layer of lithium about 1 millimeter thick on the inside of the porous wall, formed as pressurized lithium flows through the wall, would vaporize and protect the wall from melting. A larger steel pressure vessel absorbs most of the shock pressure generated by each explosion. The spray condenser absorbs the energy of the vaporized lithium. Hot lithium recirculated through a heat exchanger provides steam for a turbine. If a new pellet were ignited every second releasing 200 megajoules (MJ), the thermal power output would be about 200 Mw.

[*Source: Los Alamos Scientific Laboratory Report LA-4858-MS Vol. 1 (1972).*]

One alternative is a carbon dioxide laser that appears capable of delivering high-powered pulses in a sufficiently short time with an efficiency of 10 to 20 percent. Although the largest energy that a CO_2 laser has delivered in a 1-nanosecond pulse is 8 joules, achieved with a prototype device at Los Alamos Scientific Laboratory, Keith Boyer and his associates there are building a 1000-joule unit. It may be difficult to achieve pulses as short as a neodymium laser can produce, but Boyer is confident that 0.25-nanosecond pulses can be achieved with known techniques. He estimates that the greatest energy attainable with a single-path CO_2 laser is 2000 to 5000 joules. For a single-path laser of glass media estimates are 1000 to 2000 joules. But by using more than one laser, the total energy could be multiplied manyfold.

The greatest limitation of the CO_2 device may be that its very long wavelength radiation, 10.6 μm, will not heat a pellet efficiently. Computer simulations indicate that radiation of this wavelength will create a surplus of very hot electrons that may preheat the central pellet mass making it difficult to compress, or may simply not transfer energy to the pellet at all. No experimental studies of laser-pellet interactions have been done with CO_2 radiation yet. If shorter wavelength radiation is needed, Boyer and his co-workers believe that the best way to get it may be to multiply the frequency of a CO_2 laser rather than to use a neodymium laser. They expect that a modified CO_2 laser would still be much more efficient.

A detailed discussion of the merits of neodymium versus CO_2 lasers is currently going on among the various members of the AEC program, but it may ultimately be superfluous to the merits of laser fusion. New breakthroughs in laser technology may provide an acceptable laser at the right wavelength, whatever that turns out to be. According to John Emmett, head of the Laser Research and Development program at Livermore, breakthroughs have occurred in laser technology at the rate of almost one major innovation per year, and there are 20 to 30 systems similar to the CO_2 system that might develop into efficient lasers. One that shows great promise is a high pressure xenon laser at Livermore that emits 0.16-μm radiation and could theoretically function with 90 percent efficiency.

Few studies have been made of the engineering design for fusion reactors because so few of the criteria are established yet; the studies that have been made emphasize general features. Some problems will be common to both magnetically contained and inertially contained devices because a deuterium-tritium (D-T) mixture appears certain to be the fuel used in both. Reactor size will be determined largely by the need for a layer of lithium, 1-m thick, to absorb the high energy neutrons that will be produced in the reaction [neutrons with energies of 14.4×10^6 electron volts (14.4 Mev) will be produced in a D-T reaction, whereas neutrons from fission have an energy of only about 2 Mev].

Several engineering problems are specific to the laser-pellet fusion scheme, however. Besides a laser of great power and durability, which is certainly the foremost problem, a very strong restraining wall will be needed to contain the blast. Spherical steel vessels can be made thick enough to withstand the relatively small explosions of 10^9 joules (the equivalent of 500 pounds of TNT), but some method must be found to contain the lithium and protect the structural parts against radiation damage. A study of laser reactor feasibility carried out at Los Alamos suggests that the problem can be solved by building a double-walled enclosure with liquid lithium between the walls. If the inner wall is perforated, liquid lithium will flow through to cover its inner surface and the layer of liquid will protect it from erosion by x-rays and exploding pellet material. The reservoir of lithium (in which neutron energy is converted to heat) is large enough to assure a constant flow of heat to a steam turbine even though the fusion bursts are intermittent.

A laser reactor design that also uses liquid lithium as the primary absorbing medium for the fusion energy has been proposed by Art Fraas and colleagues at the Oak Ridge National Laboratory. The Oak Ridge concept eliminates the need for an inner wall by swirling the lithium at a velocity high enough to maintain a free-standing vortex. Fuel pellets would be injected into the vortex and ignited when they reach the center of the chamber. The Oak Ridge group also proposes to cushion the restraining vessel against blast pressure by introducing fine bubbles of inert gas into the circulating lithium. Neither the Oak Ridge nor Los Alamos reactor designs seem to be particularly well suited for multiple-beam irradiation, but presumably many variations are possible.

Both reactor designs are spare-time efforts, since the outstanding question is still

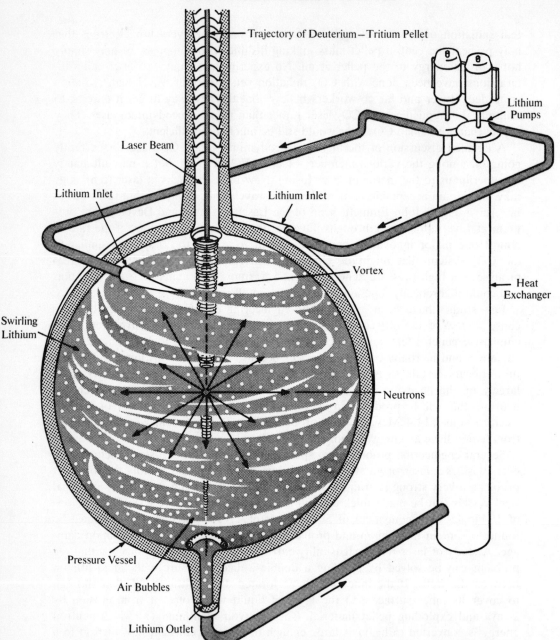

FIGURE 29. Laser fusion reactor with a lithium vortex. This design is proposed by workers at the Oak Ridge National Laboratory. There are no interior walls. The fusion energy would be absorbed in a pool of liquid lithium contained in a spherical pressure vessel some 3 to 5 meters in diameter. A free-standing vortex would be formed in the center as a result of swirling the lithium at a sufficiently high velocity. The fusion energy released in the form of x-rays and neutrons would be absorbed in the lithium, and would be removed by drawing off lithium from the bottom of the vessel, circulating it through a heat exchanger, and returning it to the pressure vessel through tangential nozzles. Gas bubbles introduced into the lithium through the perforated ring at the bottom of the vessel would attenuate the blast waves produced by the explosion of the fuel pellet.

[*Source: From "Fusion by Laser," by Moshe J. Lubin and Arthur J. Fraas. Copyright* © Scientific American, *Inc. All rights reserved.*]

Table 11. Summary of laser fusion reactor components, status, and outlook.

Component	Purpose	Principal requirements	Current status	Outlook
Laser.	Periodically provide an extremely intense burst of energy to heat a pellet of fusionable material.	Energy output = 10^5–10^6 joules; pulse width: near 1 nanosecond; overall efficiency ≥ 10 percent (laser light output to energy input); reasonable capital cost.	500 joules demonstrated; 0.1 percent demonstrated in pulsed glass lasers; 10–20 percent in CO_2 lasers which have yet to be short pulsed; low cost not foreseeable for glass lasers; conceivable for CO_2 and chemical lasers.	This development represents a large extrapolation from existing systems; a breakthrough will be required.
Light transmission system.	Provide an optically and mechanically safe channel between the laser and the pellet at its intended irradiation point.	Low optical losses; mechanically shield the laser from the blast; minimize nuclear radiation back to laser; prohibit reflection back to the laser.	Readily conceivable.	Should be a matter of straightforward development.
Pellet system.	Manufacture pellets of fusionable material (probably DT) and deliver them to the irradiation point within the containment system.	Simplicity of operation; low capital and maintenance cost; low cost per pellet.	Conceivable.	If simply solid spherical pellets are used this development should be straightforward. If complex pellet configurations are needed, the development of an economic pellet system may be difficult.
Containment system.	Provide a suitable point for irradiation of the pellet; provide blast containment of the exploding pellet; provide a neutron moderator and a means of energy removal.	Contain up to 10^{10} joule blasts; thermalize very large fraction of the neutron energy; shield all nuclear radiation (principally neutrons); operate at high temperatures to maximize system efficiency; efficiently transmit blast and neutron energy to the energy converter; low capital and maintenance costs.	Already demonstrated; technology available.	While the magnitude of this development will require considerable effort, no basic limitation appears to exist.
Energy conversion and storage system.	Convert thermal energy to electrical energy and store energy for subsequent laser pulses.	Operate at high temperatures to maximize system efficiency; low capital and maintenance costs; store energy for laser pulses efficiently and at low capital costs.	Technology available; energy storage requires development (probably of inductive energy storage).	Standard conversion techniques are readily available but more efficient techniques would require development. Inductive energy storage appears feasible and economic but is presently undeveloped.

Source: Statement of Dr. Robert L. Hirsch to a subcommittee of the Joint Committee on Atomic Energy, Nov. 10, 1971.]

whether laser fusion is feasible or not, and there is considerable difference of opinion among investigators about such basic questions as the ideal size of the pellet. Nuckolls and his co-workers at Livermore favor a millimeter pellet, releasing about 10^7 joules of energy, whereas a pellet about 1 centimeter in diameter is selected by the Los Alamos group, as well as by researchers at Oak Ridge and Rochester. It may be possible to burn more of the larger pellets, but more laser energy would be required to ignite the reaction, thus increasing the gap between the lasers available and the lasers needed. For smaller pellets, the value of the energy released would be only about 1 cent when sold as electricity. Clearly one of the problems is to produce pellets cheaply.

If both laser fusion and magnetically confined fusion should prove feasible, laser-based reactors would have several advantages. Most studies estimate that their size would be about 50 to 200 Mw of electric power, much smaller than the 500- to 2000-Mw capacity estimated for a tokamak fusion reactor. The laser system would be a suitable size for uses such as ship propulsion. Many small units could be clustered together—perhaps served by a single laser system—or small units could be widely distributed to provide electrical power for homes and industries. If many units were combined in a modular system, individual units could be shut down to replace parts damaged by radiation without significantly affecting power generation. According to Art Fraas, it appears that both the capital costs and the development costs would be lower for a laser fusion reactor.

But laser systems have been much less thoroughly studied than magnetic systems, and may seem to have fewer problems only because the concept is newer. According to some observers, the technological problems of the laser alone could prove to be as difficult as the development of an entire tokamak reactor. Building lasers with an output of 10^5 or 10^6 joules is difficult, and increasing their repetition rate to one shot per second may be even tougher. Repetition rate of the Naval Research Laboratory laser, one of the largest in the United States, is about one shot every 4 minutes! In order to keep the laser from being damaged during a pellet explosion, disposable mirrors or liquid metal mirrors have been suggested to eliminate all straight paths between the pellet and the laser, but no tests have been made.

Although the implosion concept of laser fusion is very appealing, the number of technical problems appears to be immense, even if the physics is found to be favorable. By bypassing the need for a magnetic field, the laser approach has also bypassed a long and depressing catalog of plasma instabilities that have thwarted the enthusiastic hopes of physicists for early proof of feasibility. But more than one major breakthrough per year in laser development may be necessary before scientists can make hydrogen in a reactor hotter and denser than the sun.

IV.
ENERGY TRANSMISSION

"We expect no major changes in the technology of electric power generation and transmission ... between now and 1985."

—National Economic Research Associates, Inc.

15.
ENERGY TRANSMISSION

The phrase "new energy sources" is fast becoming synonymous with "new sources of electricity." Although much work with coal is directed toward production of new types of fuels and solar energy may first be used for residential heating, most of the developments discussed in the preceding chapters are aimed at the production of electricity. Implementation of these developments will thus bring about gross changes in the pattern of energy consumption. Electricity now accounts for only about 10 percent of energy end use, but its share is expected to rise to more than 25 percent by the year 2000—a 150 percent increase in its share, but more than a 600 percent increase in consumption.

This increased reliance on electricity will raise many new questions about energy transport and storage. Should coal, for example, be burned at the mine to produce electricity that can be transmitted to centers of consumption, or should the coal be shipped to those centers for production of electricity there? Should new generating facilities be large installations located some distance from urban centers, or should they be small plants built close to areas they serve? Are the environmental and esthetic costs of electrical transmission greater or lesser than those of electrical generation? Can the cost of electricity be lowered by developing new methods for storing electricity produced during periods of low demand?

Each of the energy options might provide a different set of answers to such questions. Fossil-fuel and nuclear power plants, for example, benefit greatly from economies of scale, and both have tended to grow larger—many now under construction have a generating capacity of 1000 megawatts or more. Power plants envisioned for fusion processes that depend on magnetic containment of the plasma might be equally as large. Plants of this size would probably be centrally located.

Power plants to exploit laser-induced fusion, in contrast, could plausibly be built with generating capacities as small as 100 Mw. Most geothermal plants are also expected to be in the 100- to 200-Mw range. The optimum size for solar-thermal power plants is not yet known, but some observers believe these, too, will be relatively small. Solar and geothermal power plants will also be subjected to physical constraints: solar facilities must be built in areas that receive large amounts of sunlight, and geothermal facilities can be located only where there are geothermal resources.

Environmental restraints may also reduce the number of options available. Many nuclear power plants, for instance, will probably be built in the ocean so that seawater can be used for cooling. Also, many environmentalists are militating

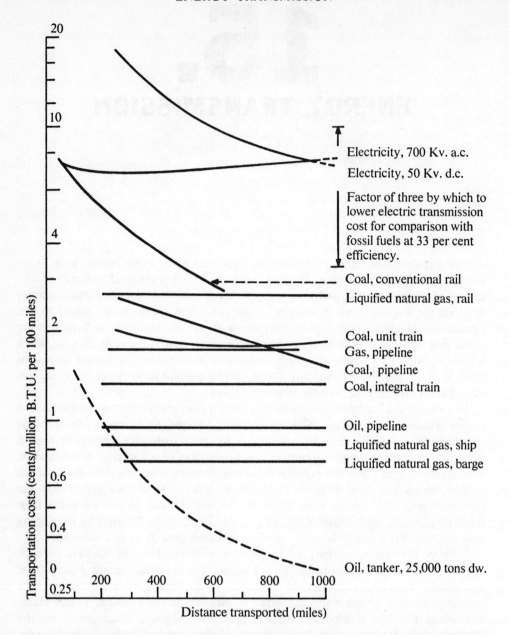

FIGURE 30. Transportation costs of different forms of energy and fuel as a function of distance. Because parallel data are difficult to acquire, and because the actual transportation cost depends critically on the particular facility as well as the particular time and place being considered, the relationships shown are only approximations.
[*Source: From "An Agenda for Energy," by Hoyt J. Hottel and J. B. Howard. Copyright* © Technology Review.]

against construction of large nuclear or fossil-fuel power plants in urban areas, against the former because of the danger of radiation leakage in an accident, the latter because of pollution from combustion of the fuels. A growing number of electrical power plants are therefore likely to be built at large distances from urban centers, necessitating an even greater amount of electrical transmission.

Electrical transmission, unfortunately, is the most expensive method of energy transportation. Direct comparisons of energy transportation costs are very difficult because of varying end uses and the lack of parallel data. But if the cost of electrical transmission by overhead lines (in cents per million Btu per mile, for example) is divided by three to compensate for the average 35 percent efficiency of conversion of fossil fuels to electricity, then electrical transmission is slightly more expensive than shipment of coal by conventional trains. This cost, however, is still about 70 percent higher than the projected costs of shipping coal by pipeline or unit train or the cost of transporting natural gas by pipeline. These costs, in turn, are about three times higher than the cost of pumping oil through a pipeline, and several times higher than that of transporting oil in large supertankers.

If the electricity is to be used to produce heat, then the relative cost of electrical transmission is approximately doubled. And if the electricity must be transmitted through underground lines, the absolute cost of transmission increases by a factor as great as 20. Clearly, then, less expensive methods of electrical transmission are desirable—perhaps even necessary—if central power stations are to be built far from urban centers.

There are three main ways to overcome the economic and environmental costs of electrical transmission. One is to increase the efficiency of electrical transmission so that much greater amounts of electricity can be carried by the same number of transmission lines. The second is to develop new, less-polluting methods for generation of electricity, such as laser-induced fusion or the fuel cell, so that small power plants can be placed near consumption centers without objections from the residents. The third alternative is to convert the electricity into another form, such as hydrogen, that can be transported more efficiently and cheaply.

Although the operating voltage of alternating current transmission lines is being pushed steadily higher and direct current transmission is now receiving limited use, the technology for all three alternatives is still embryonic. Dispersed generation of electricity by fuel cells could begin within this decade if costs can be reduced further and especially if adequate supplies of natural gas or similar fuels can be guaranteed. Limited use of superconducting transmission lines could begin during the next decade, particularly for such uses as transmission of electricity from offshore nuclear power plants to land-based distribution systems, but again, cost reductions will be a major factor in assuring widespread use. And electrical transmission via hydrogen is even further in the future, for it would require not only major technological improvements in the production and use of hydrogen, but also a massive program for adaptation of natural gas distribution systems and appliances to operate on hydrogen. It is thus unlikely that any of the new technologies could have a significant effect on electric power generation and transmission before 1985 at the very earliest.

Fuel cells and hydrogen could, however, be used for storage of electricity much sooner than for its transmission. The major benefit of this application would be a significant reduction in the amount of generating capacity required by the industry, and a corresponding reduction in capital investment.

Demand for electricity varies greatly from season to season and from day to night. Because of these variations in demand and the difficulties of storing electrical energy, the electric industry frequently operates at as little as 50 percent of its capacity. Since the generating equipment is very expensive, this unused capacity adds greatly to the cost of electricity.

The only method for storage of electricity now in use is pumped hydroelectric storage, in which off-peak power is used to pump water into a reservoir. At periods of peak demand, this water is then used to operate hydroelectric generators to produce supplemental electricity. This method of storage is cheaper than the installation of additional generators, and more than 11,000 Mw of storage capacity are now in service. The efficiency of electrical storage is only about 65 percent, however, and the number of sites suitable for hydroelectric storage is limited, so it is unlikely that this method will achieve widespread usage.

Batteries, which give storage efficiencies as high as 70 percent, are a logical alternative, but their high cost and limited lifetimes have prohibited any significant use. Claims of economic competitiveness have been made for such high-temperature systems as sodium-sulfur or lithium-chlorine batteries, but these claims have not been fully documented and it is not clear that the problems of containing these dangerous materials have been solved. Batteries may eventually find wide use, but the technology is clearly in its infancy.

An intriguing possibility presented by R. B. Korsveyer of Oak Ridge National Laboratory is storage of electrical energy as compressed air in large underground cavities. He estimates that compressed air storage would require only 10 percent of the land surface area required for pumped hydroelectric storage, and that the efficiency could be as high as 79 percent. There are, however, large uncertainties about the costs of equipment and excavation, and about the possible environmental effects, so that the potential of this method is still indeterminate.

An attractive alternative, then, would be the use of excess generating capacity for electrolytic production of hydrogen and oxygen at periods of low demand for electricity. These gases could be stored, then recombined in a fuel cell to produce electricity at times of peak demand. The overall efficiency of this process could be higher than 70 percent, and the environmental effects would be negligible. As with the other techniques, cost is the major barrier, but current work with fuel cells and electrolyzers shows promise of lowering these costs sufficiently. It is thus possible that use of such an energy storage system could provide a strong incentive for commercialization of fuel cell power plants and energy transmission via hydrogen.

16.

TRANSMISSION LINES
Three New Ways to Carry Electricity

Because the demand for energy in the form of electricity seems certain to increase drastically in the next two decades, the conventional method for transmitting it—by overhead power line—needs serious examination. If, as often predicted, U.S. power consumption multiplies six times by the year 2000, is it tolerable to install five more lines for every one that exists now, or to replace each line with a gargantuan counterpart? More than 7 million acres of land are set aside for overhead transmission. If projected demand is met with more high-voltage towers, the needed acreage will probably double. How many such towers are acceptable, particularly in scenic areas? In many large cities, where most electricity is consumed, there is simply not enough land available for enlargement of power corridors. Power lines with greatly increased capacities, particularly underground lines, appear to be needed.

Three new technologies for underground power transmission are being studied in addition to ways to improve conventional overhead lines. Transmission cables insulated with compressed gas, cables cooled to the temperature of liquid nitrogen (called cryoresistive transmission lines), and cables cooled even more, so that they become superconducting, are all being developed as alternatives to conventional underground power lines. (Conventional cables are insulated with oil and paper, and operate at ambient temperatures.)

No cryoresistive or superconducting cable has been installed commercially, but a short compressed gas cable is already being used by Consolidated Edison Company in New York City. Because the market will clearly be large, research funds are being contributed by companies that hope to sell the lines, as well as by the Atomic Energy Commission, the National Science Foundation, and the Electric Research Council, which is funded by private utility companies and the Department of the Interior. The situation now is an open race and it is still not known whether more than one of the new technologies will be needed.

The utility companies make a distinction between transmission systems, which transport electricity from generating stations to major consumption centers, and distribution systems, which carry energy to individual consumers. Lines operating at 138 kilovolts or more are usually considered to be transmission lines. Distribution lines operate at lower voltages, require much simpler technology, and are routinely being installed underground in many cities.

Transmission is almost always done with overhead lines in the United States; only about 1 percent of the transmission lines are underground, and most of these are in the major urban centers between Boston and Washington. Overhead

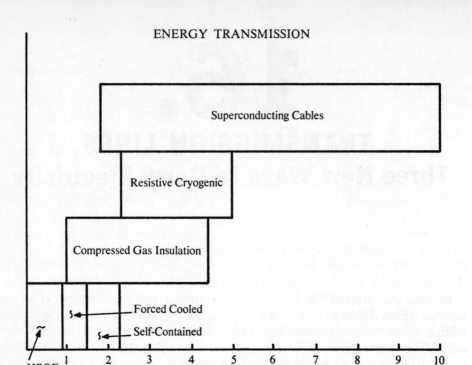

FIGURE 31. The power-carrying capabilities of new underground transmission systems. High-pressure oil-filled cable (HPOF) is of the conventional type.
[*Source: Brookhaven National Laboratory Report BNL 50325.*]

transmission lines are reliable, easy to repair, and efficient. (A 200-mile-long, 345-kv line transmits power with more than 98 percent efficiency.) Overhead lines could be designed to carry ultrahigh voltages (UHV), voltages greater than 765 kv, the highest now used in the United States. However, towers to carry a UHV cable would be more than 150 feet tall, and would require a right-of-way more than twice as wide as the present transmission lines.

Underground transmission will almost certainly replace overhead lines in situations where overhead transmission is impossible, such as to an offshore nuclear power plant, or where overhead transmission is unsafe or unsightly, such as at intersections of transmission lines with superhighways, airport runways, or other lines, or where overhead transmission is too expensive, such as in urban centers. Overhead transmission lines typically require 12 acres per mile. If the cost of land reaches $70,000 per acre, as it might in densely populated areas, the expense of land alone might be as great as the total cost of a 345-kv underground cable ($840,000 per mile). The cost of right-of-way for an underground line is usually negligible. Pipes can be laid in a 2-foot-wide trench, and the right to lay cable under the streets is often donated by the city. If land costs are not considered, however, an underground cable costs 6 to 20 times more than an overhead cable with the same power-carrying capacity.

Underground cables now available often have too low a transmission capacity to replace overhead lines. Underground lines are limited by earth's inability to absorb heat produced in the cable. For a conventional underground cable designed to carry 230 kv the maximum power capacity is about 300 million volt-amperes (MVA), even if the trench is packed with a filler of high thermal conductivity. (About 300 MVA is the power needed by a city of 200,000.) In order to transmit the amount of power now routinely generated with steam turbines, many such cables would be needed. For instance, in Bratsk, Siberia, Soviet engineers have built an underground system of two raceways, each holding six oil-filled cables. The total capacity is 2700 MVA. In the Soviet system the cables are cooled with forced air, but even more efficient methods of cooling, such as recirculating the oil through a heat exchanger, will be necessary. Even with improved cooling, oil-filled cables will probably have only twice their present capacity.

The advantage of the newer cable technologies is that heating of the cable can either be reduced, in the case of compressed gas cables, or effectively controlled, in cryoresistive and superconducting cables. The benefit is not that a greater fraction of the power reaches the other end but that much more power can be carried without exceeding the tolerances of safety. For example, in a cryoresistive line the resistance of the conductor is reduced by a factor of 10 when the line is cooled to the temperature of liquid nitrogen ($77°K$). However, all refrigerators become much less efficient at low temperatures. A typical refrigerator operating at $77°K$ would use about 9.5 watts of power to pump 1 watt of heat away from the cable. Thus the savings in power dissipation due to a lower resistance are canceled by the inefficiency of a very-low-temperature refrigerator.

A similar effect occurs with superconducting transmission lines. The resistance of a superconducting line is zero for direct currents (d-c); but for alternating currents (a-c) the magnetic field from the current produces heat which must be pumped away with a refrigerator. The total loss in an a-c superconducting cable, after the inefficiency of the refrigerator is taken into account, will probably be

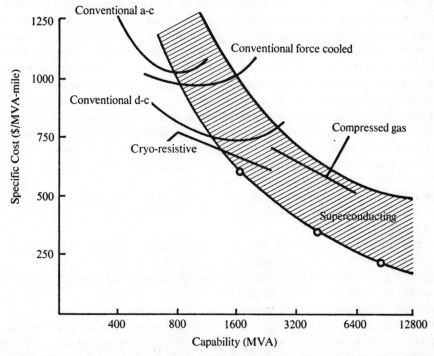

FIGURE 32. Costs of various types of underground transmission lines. The cost of high-capacity overhead lines is about $100 per MVA per mile (based on the 1971 installation of a 765-kv line with 3000-MVA capacity by the American Electric Power Company). [*Source: Brookhaven National Laboratory Report BNL 50325.*]

about the same as the losses in an overhead transmission line operating at ambient temperature (but certainly less than the losses in the present underground cables).

Figure 32 indicates the power capabilities of conventional oil-filled cables and the three alternatives. The maximum capacity of the conventional cable, about 1500 MVA, is clearly exceeded by the others. The superconducting cable has the greatest potential for carrying large amounts of power. For most designs, it has been estimated that a single superconducting cable could carry more than 10,000 MVA, the power needed for all of New York City.

Underground cables are not without severe limitations. The configuration of an underground line (namely, an arrangement of two concentric conductors) introduces a capacitance that is very much larger than that of overhead lines. The capacitance induces a current (called a charging current) that does not contribute any effective power, but nonetheless produces large heat losses. The capacitance of conventional cables is particularly great because of the choice of insulation and because they are made very compactly. As a result, conventional underground cables have a large charging current, which limits the usable length to about 20 miles.

Compressed gas cables have much lower charging currents and therefore can be used to carry more power for distances of over 200 miles. Because of the superior electrical properties of sulfur hexafluoride, the gas used as an insulator, heat produced is significantly less than that from a conventional cable. However, compressed gas cables have several disadvantages—not the least of which is cost—that have limited the longest installation so far to 1300 feet. For example, a 500-kv compressed gas cable capable of carrying more than 2000 MVA of power would cost about $2 million per mile, according to H. C. Doepken of the High Voltage Power Corporation, Westboro, Massachusetts. Each of the three phases of a 500-kv line would be a rather large, rigid aluminum pipe 20 inches in diameter. Individual sections 40 feet long would be welded together on-site, a time-consuming and costly process. Because the three pipes are laid side by side, compressed gas lines require very wide trenches. A 500-kv line, for example, would require an 8-foot-wide trench. Some utility officials think that the size of the trenches alone may prohibit use of compressed gas cables in many urban areas. However, the great advantage of gas cables is that they are available today. Doepken and others think that it will soon be possible to double the capacities of compressed gas lines simply by cooling the outer cable jackets with water. High Voltage Power Corporation and other companies are trying to develop a compressed gas cable with all three phases in a single jacket, but whether the trench size needed for a line of very high capacity could be reduced to something less than the width of a city street is still to be determined.

A gas-insulated transmission line somewhat different from the compressed gas lines now in use has been proposed by Benjamin M. Johnson of the Battelle Pacific Northwest Laboratories, Richland, Washington. If an insulating gas other than sulfur hexafluoride were used, it could be circulated into the cable in the liquid state, then evaporated by the heat produced in the cable, and finally recondensed into liquid in an external cooling unit. The system would require cooling units at frequent intervals, as would the cryoresistive and superconducting cables, but it is likely that no new technological development would be necessary. The Battelle proposal is not yet being commercially developed, however.

Cryoresistive transmission lines would also solve the power limitations of conventional cables by cooling the system with external refrigerators, but the pro-

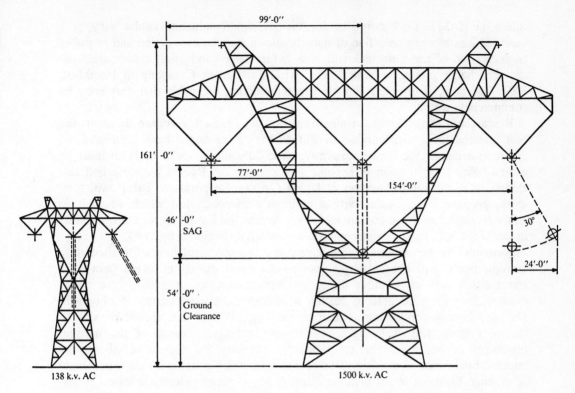

FIGURE 33. New overhead transmission towers. A drawing indicates to scale how the sizes of support towers for overhead transmission lines must be increased to carry more power. Transmission towers carrying power at 138 kv are common today, but towers rated for 1500 kv may be needed by 1985.
[*Source: C. H. Haines, Commonwealth Associates, Inc. Jackson, Michigan.*]

ponents of the cryoresistive technology hope that by avoiding the extremely cold temperatures needed for superconductivity they can produce a working transmission line sooner and with less research and development. At General Electric Company in Schenectady, New York, a 500-kv cryoresistive cable is now undergoing laboratory tests. It is a flexible cable with aluminum conductors, which will be cooled to the temperature of liquid nitrogen and insulated with a synthetic paper (Tyvek). It is expected that refrigerators will be installed every 7 miles, so the cooling network will be about as extensive and complex as the electrical network. The capital costs of the refrigerating system are certain to be quite large, and the reliability of the system will have to be proven through an extensive program of testing. Careful selection of the locations of the initial commercial installations will also be necessary.

Superconducting transmission lines will be made of metals or alloys that have no resistive losses at very low temperatures. Because these metals must be cooled to extremely low temperatures in order to become superconducting, problems of refrigeration and thermal insulation will certainly be greater than for cryoresistive lines. However, superconducting cables could carry very large amounts of power, and the cables would be very compact. In some estimates the diameter of a 10,000-MVA cable is as small as 60 cm. The cost of a superconducting cable would certainly be immense, but it would transmit so much power that the cost per kilowatt is projected to be less than the cost of conventional underground cables (Fig. 33). Cost estimates for various superconducting cables vary by a factor of 2 or more, a reflection of uncertainties about the best design and probably an indication of the time that will pass before superconducting cables are well-tested, reliable products. Nonetheless, AEG-Telefunken Company in Frankfurt, West Germany, is testing a flexible superconducting d-c cable that may soon be commercially available.

Researchers in the United States have not yet reached a consensus about the best material for a superconductor. But all the cables tested have been made of niobium, a metal that is rare, expensive, and difficult to work. Niobium must be cooled below $9°K$ in order to become superconducting. Robert W. Meyerhoff and associates at the Linde Division of Union Carbide Corporation, Tarrytown, New York, propose a rigid cable with a niobium superconductor, which would have either liquid helium (the coolant) or a vacuum as the insulating medium. According to Meyerhoff, such a cable could be commercially available by 1984.

Niobium has the advantage of ductility and low electrical losses for alternating current, but it will not carry the increased current during a severe fault in a transmission network without losing its superconducting properties (a-c transmission lines must be able to sustain a tenfold current overload). Niobium-tin (Nb_3Sn) is superconducting at temperatures up to $18°K$ and should be able to sustain a fault without losing transmission capability. Because of the unusual properties of helium at the temperatures necessary for superconductivity, the improved efficiency of a refrigerator operating at the higher optimum temperature of niobium-tin would more than compensate for its greater electrical losses. A disadvantage of niobium-tin is that its physical properties, such as ductility, are poorer than those of niobium.

Niobium-tin superconducting transmission lines are being studied at the Brookhaven National Laboratory, Upton, New York, by Eric Forsyth and associates and at Stanford University, Palo Alto, California, by Theodore Geballe. The Brookhaven study, which is supported by the National Science Foundation under the Research Applied to National Needs (RANN) program, proposes a flexible cable with a helical conductor and a solid insulator. The Stanford design is similar, but the superconductor is fabricated in 50 or 100 thin layers rather than in 3 or 4 layers, as in the Brookhaven design.

Direct current superconducting transmission lines are also being studied as a possibility for power transmission in the future. Direct current is advantageous for underground transmission because a line will generally have twice the capacity for d-c current as it does for a-c current. However, an expensive convertor is necessary at each end of a d-c line in order to couple it into the existing a-c network. Direct current lines have seldom been used in the United States (although more often than in Europe). For very long transmission lines, however, the savings in the cost per mile of the line will more than compensate for the cost of having convertors at both ends (about $30 per thousand watts of transmission capacity). The Bonneville Power Administration has an 850-mile-long d-c overhead transmission line from Oregon to Southern California.

Direct current transmission by superconducting lines would involve no electrical losses at all, so the refrigerators would only have to compensate for the relatively small amount of heat that leaks into the cable because of inefficient thermal insulation. However, the electrical losses in convertors would be about 1.25 percent at each end, an amount comparable to heat losses in a 200-mile a-c overhead line. E. F. Hammel and associates at the Los Alamos Scientific Laboratory,

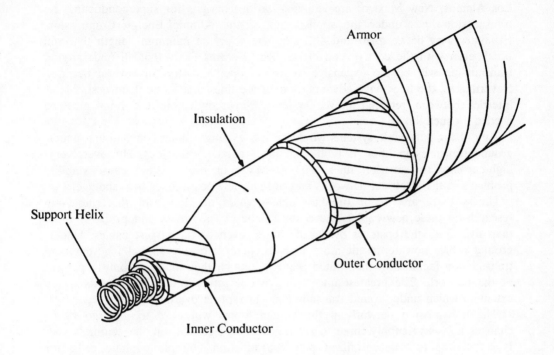

Armor

Insulation

Support Helix

Outer Conductor

Inner Conductor

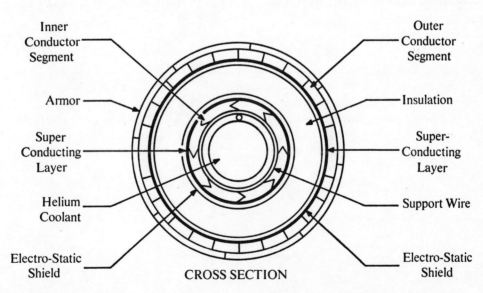

Inner Conductor Segment

Outer Conductor Segment

Armor

Insulation

Super Conducting Layer

Super-Conducting Layer

Helium Coolant

Support Wire

Electro-Static Shield

Electro-Static Shield

CROSS SECTION

FIGURE 34. A proposed design for a flexible superconducting underground transmission line. Superconducting material is deposited on the surface of copper or aluminum segments, which are helically wound to produce the inner and outer conductors. The insulation would be wrapped in plastic tape, thin enough for ease of bending the cable. The cable is cooled from inside the inner conductor with liquid helium. Three such cables enclosed in a larger pipe would be needed to carry a-c power.
[*Source: Brookhaven National Laboratory Report 50323.*]

Los Alamos, New Mexico, are studying various designs for superconducting d-c transmission lines under the sponsorship of the Atomic Energy Commission. Estimates vary between 50 and 400 kilometers for the minimum length that will break even with the cost of convertors. But it seems likely that all underground transmission will be too expensive to use in open country for several decades. Nevertheless, d-c lines have advantages over a-c lines that make them desirable as interties between regional power systems. Superconducting d-c lines may be needed for such links.

Any of the new cable systems would be less expensive than conventional underground transmission lines, if built with large enough capacities. However, very-high-capacity transmission lines, 10,000 MVA or more, could cause stability problems in the electrical networks and present new possibilities for sabotage.

Unless there are major breakthroughs—which are not now discernible—in research on these new cables, the costs will be at least three and probably more than five times the costs of overhead cables, even of the largest cables. Underground cables now represent more than 10 percent of the capital investment in transmission facilities in the United States, although they make up only 1 percent of the network. The greatest uncertainty about future transmission costs is the extent to which underground transmission will replace overhead lines.

The transmission network in the United States will have to undergo major changes if it is to supply three to six times more electricity at the century's end. It is possible that decentralized power production (by photovoltaic cells, for instance) could alleviate the need for more electrical transmission or that cheaper means of energy transmission (such as piped hydrogen) might supplant the electrical transmission line. But for the next decade or two, it appears that the rising costs of electrical power transmission will add to the increasing price of raw energy.

17.

FUEL CELLS
Dispersed Generation of Electricity

The principle of the fuel cell was discovered by Sir William Grove in 1839, but it remained little more than a scientific curiosity until the first practical fuel cell was demonstrated in 1959 by Francis T. Bacon and J. C. Frost of Cambridge University. Since that demonstration, fuel cells have been widely used in the space program, but their high cost has effectively precluded their use as earthbound power sources. Only recently has it begun to seem likely that the cost problems could be overcome and that fuel cells could be commercially viable within this decade.

The road to viability has been a strange one. The euphoria of the space program attracted a number of companies to fuel cell development, but disillusionment set in rapidly. It is comparatively easy to produce electricity efficiently and for long periods of time when money is no object; it is far harder, they found, to do it when that electricity must compete economically with the relatively cheap product of large commercial generators.

Government research funds, furthermore, were granted almost exclusively for space and military applications, and even those funds dropped from nearly $16 million in 1963 to about $3 million in 1970. Unwilling or unable to assume the substantial investment required for commercialization, companies that had so eagerly rushed into fuel cell development quietly abandoned their research programs or reduced them to token operations. Only one company is now pursuing a full-scale commercial fuel cell program—the Pratt & Whitney Aircraft division of United Aircraft Corporation, East Hartford, Connecticut.

Pratt & Whitney is, however, supported by the natural gas industry, which views fuel cells as an attractive, environmentally sound way to obtain a premium rate of return on natural gas sales by upgrading the gas to electricity. The electric industry has also provided support because fuel cells promise to be small, clean power sources that can be installed quickly throughout its distribution systems to supplement central power stations without objections from residents or ecologists.

To date, 43 U.S. and three foreign utilities and Pratt & Whitney have invested more than $50 million to prove the technical feasibility of commercial fuel cells. They will probably invest twice that amount during the next 3 years in an attempt to demonstrate the economic viability of fuel cells.

The object of this effort embodies a relatively simple concept. The fuel cell's main components are a fuel electrode (anode), an oxidant or air electrode (cathode), and an electrolyte. In a typical application, hydrogen is fed to the anode, where it is catalytically converted to hydrogen ions, releasing electrons to an external circuit. At the cathode, these electrons reduce oxygen to ions which then

Table 12. Federal standards for maximum emission of air pollutants from central generating stations (in kilograms of pollutants per 1000 kilowatt-hours) compared to maximum emissions from experimental fuel cells.

Pollutant	Gas-fired utility	Oil-fired utility	Coal-fired utility	Experimental fuel cells
Sulfur dioxide	nr	3.43	5.09	0.000121
Oxides of nitrogen	0.915	1.29	2.96	0.110
Hydrocarbons	nr	nr	nr	0.105
Particulates	0.457	0.43	0.424	0.000014

Federal emissions standards effective 8–17–71. nr = no requirement.
[Source: Pratt & Whitney Aircraft]

migrate through the electrolyte and combine with hydrogen ions to form water. This process—effectively the reverse of electrolysis—continues as long as fuel and air are supplied to the cell, and its structural integrity is maintained. Fuel cells can be adapted to a variety of fuels by changing the catalyst, but hydrogen fuel cells are the most efficient and the most highly developed.

A single hydrogen fuel cell generates from 100 to 200 milliamperes of direct current per square centimeter of electrode and a potential of about 1 volt. By connecting a number of cells, it is possible to create useful potentials of 100 to 1000 volts and power levels of 1 kilowatt to nearly 100 megawatts.

A fuel cell power plant generally also contains a reformer and an inverter. The reformer uses chemical processes to convert the fuel to a form that can be utilized by the cell. In some Pratt & Whitney systems, steam and a catalyst are used to convert hydrocarbons into hydrogen and carbon dioxide. The inverter changes the direct current output into alternating current at the frequency and voltage levels required.

Some of the advantages of fuel cells over conventional power sources are obvious. Emission of air pollutants, for example, is negligible because fuel cell operation is not based on combustion. Thermal pollution of waterways is not a problem because excess heat is released directly to the atmosphere. Noise is also low because only the fuel and cooling systems have moving parts.

Another major advantage is efficiency of electrical production at low power. The best steam and gas turbine generators now operate at a thermal efficiency of about 39 percent at power levels greater than 100 Mw. The best fuel cell power plants may be only slightly more efficient, but that efficiency can be achieved at power levels as low as 25 kw, where other generating systems are much less efficient (Fig. 37). If the power plant is operated on hydrogen and doesn't require a reformer, the efficiency may be close to 55 percent. Conventional generators, moreover, are much less efficient when operated at less than maximum capacity, whereas fuel cells maintain high efficiency even when operated under a partial load.

The primary problems of fuel cell development are initial cost and service life. Pratt & Whitney has estimated that current technology would permit the production of fuel cell power plants, independent of size, at a cost of about $350 to $450 per kilowatt and with a service life of about 16,000 hours. To be competitive commercially, the company says, the cost should be halved and the service life doubled. These objectives must be met by increasing the power output and by developing less expensive, more stable construction materials.

Much progress has apparently been made in these areas, but Pratt & Whitney's

SINGLE FUEL CELL

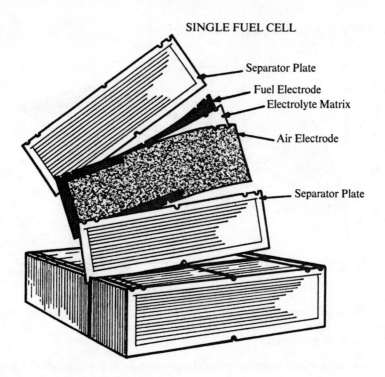

FIGURE 35. A typical fuel cell assembly for a prototype commercial fuel cell.
[*Source: Pratt & Whitney Aircraft*.]

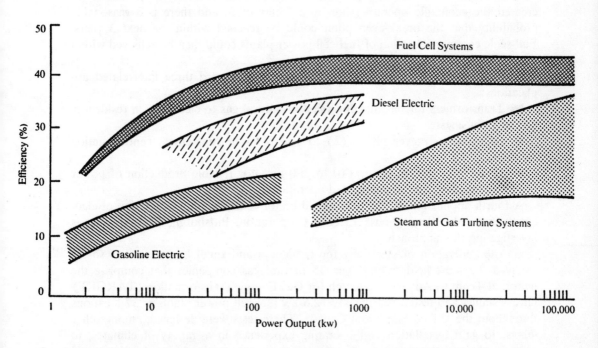

FIGURE 36. Thermal efficiency of electrical production of fuel cells and other types of generators as a function of output.
[*Source: Pratt & Whitney Aircraft*.]

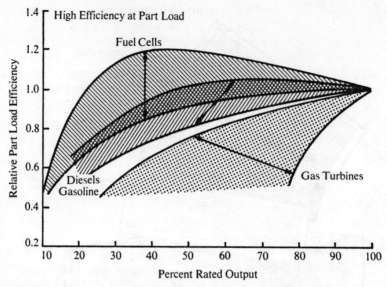

FIGURE 37. Relative thermal efficiency of electrical production of fuel cells and other types of generators when operating at less than their rated capacity. [*Source: Pratt & Whitney Aircraft.*]

statements on the subject must be taken largely on faith because all the new technology and materials developed in its program are proprietary. More than 85 percent of the materials and technology on which the power plants are based, the company claims, have been developed within the past 5 years. These improvements have already reduced the cost per kilowatt by a factor of nearly 20 and have increased the economic operating life by a factor of 5, and there is a reasonable probability that the break-even point could be reached within the next 3 years. Full-scale commercialization of fuel cell power plants could then be achieved within this decade.

Pratt & Whitney's efforts are directed primarily toward three interrelated applications:

► Transformers for on-site conversion of natural gas to electricity in residences and small businesses.

► High output power plants (25 to 100 Mw) to supplement central station facilities.

► Low output power plants (10 to 200 kw) for reliable production of power in remote locations and in unattended operation.

A fourth application of fuel cells would be for central station generation of electricity, but only Westinghouse Electric Corporation, Pittsburgh, Pennsylvania, is investigating this approach.

On-site generation of electricity for residences and small businesses has so far received the most field testing. The 35 natural gas companies that comprise the nonprofit Team to Advance Research for Gas Energy Transformation (TARGET) have installed and tested nearly sixty 12.5-kw fuel cell power plants at 37 locations throughout the United States and Canada. These tests were designed, among other things, to gain installation and operating experience in a variety of climates, to assess problems related to the ratio of peak to average demands for electricity, and to obtain experience in estimating the maximum demand in different types of buildings. Developing the capability to determine the customer's precise electrical requirements, Pratt & Whitney says, is as important for commercial success of fuel cell energy service as is development of the power plant itself.

The goal of the TARGET program is to produce electricity at a cost competitive with that of central station generation—with no credits for environmental benefits or waste heat utilization. One major factor that may allow attainment of this goal is the relatively high price of transmitting and distributing electricity. Transmission and distribution costs for natural gas are only about 20 to 30 percent of those for electricity, per unit of energy. Capital investment should also be significantly lower for fuel cell energy service, the company predicts.

Another factor is lowered fuel use. While a central generating station may have a thermal efficiency of 39 percent, transmission losses reduce the overall system efficiency to about 34 percent. Utilities, moreover, must supplement their main generators with less efficient standby and peak power sources. The net effect, TARGET assumes, is that on-site fuel cell power plants produce about 20 to 25 percent more useful energy per unit of fuel.

A final benefit of on-site fuel cells, and one that has not yet been fully explored, is the use of by-product heat from the power plant. Pratt & Whitney suggests that this energy could not only help provide heating, hot water, and air conditioning, but could also be used to purify, deodorize, humidify, or dehumidify air, to dispose of waste, and to purify water. Such uses would make the overall thermal efficiency even higher.

The principal argument against natural gas fuel cells would seem to be the shortage of natural gas. Since gas supplies are already tight, new uses would presumably only aggravate the problem. But fuel cells, argues TARGET president Robert Suttle, are the most efficient method of converting fossil fuels to electricity. If natural gas can be used more efficiently in this fashion, he points out, there will be greater quantities of oil and coal available for other applications, the implication being that natural gas may eventually be diverted from industry and utilities to residential use.

Most of the arguments favoring on-site fuel cells can also be applied to the use of fuel cells by electric utilities. The attributes that particularly suit fuel cells for this use, however, are their ability to be placed virtually anywhere without disruption of the environment and the relatively short time required for their construction. Electric utilities could thus use such plants to supplement their systems in rapidly growing areas and in areas where siting or transmission problems prevent the construction of conventional plants.

Fuel cells could also be useful as part of a storage system for electricity to supplement the generating capacity of nuclear or thermonuclear power systems. During periods of minimum demand for electricity, the output of central stations could be used for electrolysis of water; at times of peak demand, the hydrogen thus produced could be converted to electricity at high efficiency, thereby lowering the total capacity requirement of the system.

The larger fuel cell power plants would, in essence, be composed of several smaller units linked together. This type of modular construction would speed installation by allowing units to be assembled at a central plant and trucked to the site. It also provides an inherent redundancy that would allow generation of electricity even when some sections of the facility are disabled or shut down for maintenance.

The chief difference from residential units is that the larger facilities will most likely consume a distillate fuel such as No. 2 fuel oil or jet fuel. This choice is

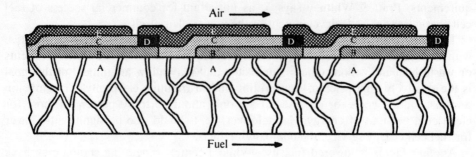

FIGURE 38. Cross-section of a high temperature fuel cell assembly. Key: (A) porous carrier tube; (B) fuel electrode; (C) electrolyte; (D) interconnections; (E) air electrode. [*Source: Westinghouse Electric Corporation.*]

dictated by the expected shortage of natural gas and the absence of practical technology for reforming heavier hydrocarbons. Such fuels are relatively expensive, but they are already widely used for supplemental generators; the higher efficiency of fuel cells could thus offset the high cost of these fuels. The first field test of such a unit began in late 1972 when a 37.5-kw experimental unit was installed at Public Service Electric and Gas Company in Newark, New Jersey.

Power plants for use in remote locations represent a much smaller market than those previously discussed, but the technical requirements are similar. The chief attributes of fuel cells for this application are unattended operation, freedom from maintenance, and quiet operation. Engelhard Minerals and Chemicals Corporation, Murray Hill, New Jersey, is developing fuel cells that consume ammonia or methanol for this market, but it too does not yet have a commercial product.

Central-station generation of electricity with conventional fuel cells is impractical. Because such cells are no more efficient than the best large turbines, no advantage is gained in their use. Westinghouse has been investigating high temperature fuel cells to be used in conjunction with coal gasification plants, and these could have thermal efficiencies greater than 60 percent.

At temperatures near 1000°C, oxidation at the anode is spontaneous and requires no catalyst (conventional catalytic fuel cells operate at about 250°C). The Westinghouse cells, which operate on carbon monoxide, can generate as much as 600 milliamperes of direct current per square centimeter of electrode. The main problem is the high cost of materials that resist interaction at high temperatures.

Arnold O. Isenberg of the Westinghouse Research and Development Center is investigating a series of metal oxides for use in such cells. The solid electrolyte is zirconium oxide doped with yttrium oxide. The yttrium stabilizes zirconium oxide in a cubic fluorite crystal structure and creates oxygen ion vacancies; these vacancies permit oxygen ion conduction at elevated temperatures. The fuel electrode is finely divided nickel in a zirconium oxide matrix. The air electrode is antimony-doped stannic oxide or tin-doped indium oxide.

The cells are mounted on a porous, calcium oxide-stabilized zirconium oxide tube that serves both as a cell carrier and as a conduit for the fuel. A typical tube has an inside diameter of about 1 centimeter and a wall thickness of about 1.5 millimeters. The fuel electrode is sintered to the carrier, and the electrolyte and air electrode are then applied by chemical vapor deposition to give the structure shown in Fig. 38. Each layer is about 30 micrometers thick. The cells are electrically connected in series with a gas-tight layer of chromium oxide, also applied by chemical vapor deposition.

In operation, a mixture of carbon monoxide and a small percentage of hydrogen is fed through the core of the tube. Electrochemical oxidation at the fuel electrode converts them to carbon dioxide and water vapor, thereby releasing electrons to an external circuit. Air is forced past the outside of the tube to provide oxygen and to remove excess heat (which can, in principle, be used to maintain the operating temperature in a coal gasifier). A 120-centimeter tube of this design, containing about 120 cells, would generate more than 80 watts, but the best that Isenberg has obtained so far is a 15-cell unit that generates 8 watts.

The foremost difficulty in constructing such assemblies is applying the components in a thin film that is pinhole and crack free; if fuel leaks through the cell assembly, it will burn in air and destroy the electrodes. The chromium oxide interconnections also work poorly, and may be replaced with other metal oxides.

A commercial fuel cell of this design is obviously some distance in the future, but a Westinghouse contract with the Department of the Interior's Office of Coal Research calls for delivery of two 100-watt units within 2 years to show progress in process and materials development. A 1-kw fuel cell could be produced within 4 years, Isenberg suggests, and a 100-kw prototype power plant within 10 years. Full-scale commercialization of high temperature fuel cells, however, seems certain to be many years beyond introduction of the Pratt & Whitney units.

18.

HYDROGEN
Synthetic Fuel of the Future

It may take 50 years, 100 years, or longer, but the time is approaching when gas, oil, and coal will no longer be available for use as fuels. Possibly, reserves of these fuels will be depleted by then. Probably, production will not be able to keep pace with demand (Fig. 39). But most likely, the remaining reserves will become far too valuable as feedstocks for chemical production to be burned simply for their energy content.

By that time, of course, nuclear fission and fusion—and perhaps solar energy—will almost certainly be the major energy sources and should be capable of supplying all our energy needs. Most developmental work on these sources has emphasized the production of electricity, however, while only about 10 percent of energy end use is supplied by electricity. The remainder is supplied by the combustion of fuels to produce heat that is used in industry, homes, and transportation. It is likely both that electricity will play a larger role in supplying future energy demands and that heat from nuclear reactors will be utilized in large nuclear/industrial complexes, or nuplexes. Nonetheless, there will remain a strong demand for portable, fluid fuels, particularly for applications in transportation, and the most likely response to this demand will be a vastly increased production of hydrogen.

Hydrogen, of course, is not an alternative primary energy source, because large amounts of energy are required to produce it. Rather, it holds promise of being a highly efficient energy carrier that can be used in situations where transfer of energy as electricity is inefficient, impractical, or impossible. It is this potential that has generated such widespread interest in the possibility of a "hydrogen economy."

In many ways, hydrogen is an ideal fuel. When it is burned in air, the only possible pollutants are nitrogen oxides derived from the air itself, and concentrations of these are generally lower than concentrations produced by other fuels. When it is burned in pure oxygen, the only product is water and there are no pollutants at all. The ignition energy of hydrogen is about 0.02 millijoule, less than 7 percent that of natural gas, so it can readily be used in low-temperature catalytic burners that also produce no pollutants.

The energy content of hydrogen gas is 325 Btu per standard cubic foot (scf), less than a third that of natural gas. The lower viscosity of hydrogen permits a three-fold increase in flow capacity of a pipeline, however, and one pipeline can carry nearly equal energy contents of either fuel—although hydrogen transmission requires a greater pumping capacity. The energy content per unit mass of liquid hydrogen is about 2.75 times greater than that of hydrocarbon fuels, so that it is an ideal fuel for rockets and airplanes. The volume of liquid hydrogen is much

117

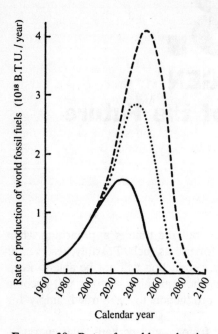

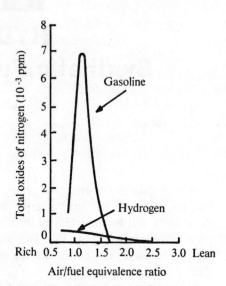

FIGURE 39. Rate of world production of fossil fuels if total reserves are assumed to be 85×10^{18} Btu (solid line), 155×10^{18} Btu (dotted line), or 226×10^{18} Btu (dashed line). [*Source: Martin A. Elliott, Texas Eastern Transmission Corporation.*]

FIGURE 40. Emissions of nitrogen oxides (in parts per million) from single-cylinder laboratory engines operating on gasoline and hydrogen. [*Source: Roger J. Schoeppel, Oklahoma State University.*]

greater than that of a comparable weight of hydrocarbons, however, since its specific gravity is only 0.07.

There is some concern—often called the "*Hindenburg* syndrome"—about the safety of hydrogen, but most such fears seem exaggerated. Hydrogen has a higher diffusivity than natural gas, so that it leaks faster, but this diffusivity also allows it to dissipate faster. Hydrogen also has a wider range of explosive concentrations in air than does natural gas, but the lower explosive limit is the crucial one, and this limit is nearly the same for both gases. Large quantities of hydrogen are shipped and used, both in this country and in Europe, with a high degree of safety.

Because it is portable and essentially nonpolluting, the most obvious application of hydrogen is in transportation. Several investigators, such as K. V. Kordesch of Union Carbide Corporation's Parma Research Laboratory, Cleveland, Ohio, have used hydrogen fuel cells to power automobiles. A simpler approach, however, is to burn hydrogen in an internal combustion engine, and the modifications required are generally quite small.

Roger J. Schoeppel of Oklahoma State University, Stillwater, has shown that the torque, power, and efficiency of hydrogen-fueled engines are comparable to those of gasoline-powered engines. Nitrogen oxides are the only pollutants, and the amounts produced are much lower than in conventional engines (Fig. 40). Robert R. Adt, Jr., of the University of Miami, Coral Gables, Florida, has also demonstrated the conversion of automobile engines to run on hydrogen. Adt suggests that the efficiency of such engines can be as much as 50 percent greater that that of gasoline engines.

The problems of conversion are so minor, in fact, that the major impediment to hydrogen use is storing enough fuel. Gaseous hydrogen, even when highly pressurized, is far too bulky and requires containers too heavy to give a reasonable operating range; liquid hydrogen requires a costly, potentially dangerous cryogenic storage tank.

One approach to this problem has been presented by Harold Sorenson of International Materials Corporation, Boston, Massachusetts. IMC has developed a compact conversion system that reforms unleaded gasoline into hydrogen and carbon dioxide by catalytic cracking. Fuel is thus stored in a conventional gasoline tank, and hydrogen is produced only as it is needed. But such a system is useful only while there are adequate supplies of hydrocarbon fuels, and is thus only a temporary solution.

One promising alternative is storage in the form of metal hydrides. Because of its small molecular size and high diffusivity, gaseous hydrogen is able to penetrate the lattice structure of solid metals or alloys and bind at various sites in the unit cell of the crystal. (This process can, unfortunately, also occur in a hydrogen pipeline, embrittling the metal and producing structural fatigue.) For many metals, such as titanium, the penetration is so great that the concentration of hydrogen per unit volume is actually greater than in liquid hydrogen. The hydrides are formed simply by exposing the metal to pressurized hydrogen. Hydride formation is exothermic and can be reversed by the application of heat; waste heat from the combustion process can thus be used to free the hydrogen.

Philips Research Laboratories, Eindhoven, the Netherlands, has found that hydrogen can be stored efficiently by intermetallic compounds of the type AB_5H_r, where A is a lanthanide rare earth element, B is nickel or cobalt, and x can be as large as 6. R. H. Wiswall, Jr., and J. J. Reilly of Brookhaven National Laboratory, Upton, New York, have obtained similar results, and have also found that the hydrides of vanadium, niobium, and their alloys are efficient in storing hydrogen. Each of these systems, however, suffers from the high cost of the metals. Less expensive metals will probably be required to make hydride storage practical.

Cryogenic storage could be used for subsonic or supersonic aircraft, and might even be necessary for hypersonic aircraft, according to Robert D. Witcofski of the National Aeronautics and Space Administration's Langley Research Center. The greater energy content of liquid hydrogen would more than compensate for the increased weight of cryogenic storage tanks, he says, and would provide superior range or payload capabilities. Liquid hydrogen, moreover, has more than 30 times the heat-sink capacity of conventional jet fuel, and could thus be used for cooling the aircraft surfaces at hypersonic speeds. Emissions of nitrogen oxides from such

Table 13. Comparative fuel properties of hydrogen and gasoline (A) or butane (B).

Property	Hydro-gen	Other fuel
Heating value, Btu/lb	53,500	20,000 (A)
Minimum ignition temperature, °F	1065	1000 (B)
Theoretical flame temperature in air, °F	3887	3615 (B)
Flammability limits, % by volume in air	4.0–74.2	1.9–8.6 (B)
Maximum flame velocity, ft/sec	9.3	1.03 (B)
Specific volume liquid, liters/kg	14.3	1.33 (A)

[Source: Roger J. Schoeppel, Oklahoma State University]

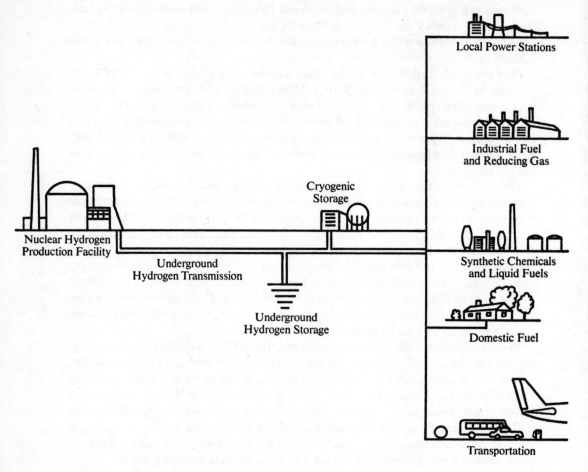

FIGURE 41. The hydrogen economy.

planes, he contends, would be less than 25 percent of those projected for a conventional supersonic transport.

Many problems must be solved before such aircraft could be built. Foremost among them is obtaining liquid hydrogen at a low enough cost (5 to 10 cents per pound, compared to today's price of about 45 to 60 cents per lb). Other problems include insulation of the storage tanks and the development of construction materials that retain their strength despite gross temperature variations.

Hydrogen could also be used to transmit energy over long distances. Transmission and distribution now account for about 45 percent of the delivered cost of electricity, about $2.22 out of a total cost of $4.89 per million Btu. That share will rise sharply as environmental considerations mandate more use of underground transmission lines.

Transmission and distribution costs for natural gas, in comparison, average only about 47 cents per million Btu; similar costs for hydrogen might be 20 to 120 percent higher. Part of the price advantage is also nullified by inefficiencies in converting electricity to hydrogen. Even with currently available technology, nonetheless, Derek P. Gregory of the Institute of Gas Technology, Chicago, Illinois, calculates that energy transmission by hydrogen is less expensive than transmission by electricity at distances greater than 400 kilometers if the energy is to be used to produce heat. If underground electrical transmission lines are required, hydrogen is less expensive at distances of only 32 km.

Another advantage of hydrogen use is increased efficiency of the power plant. Because of large time variations in demand for electricity and the great difficulties of storing electrical energy, the electric industry frequently operates at as little as 50 percent of installed capacity. If surplus electricity were converted to hydrogen and stored, the industry could operate near 100 percent of its capacity at all times, and the stored hydrogen could be used to meet peak demands for power.

Widespread use of hydrogen for energy distribution will probably depend on two factors—increased efficiency of conversion of hydrogen to electricity in fuel cells (discussed in Chapter 17), and increased efficiency of conversion of nuclear or perhaps solar energy to hydrogen. At least four techniques for production of hydrogen from water are generally considered: electrolysis, endothermic chemical decomposition, photolysis, and bioconversion. (More than half of the hydrogen used today is produced by catalytic cracking of hydrocarbons, but this route is impractical if hydrogen is to be a hydrocarbon substitute.) Only electrolysis has been proved in practice.

Commercial electrolyzers typically operate at efficiencies of about 60 to 70 percent—although some prototypes have reached 85 percent. Gregory points out that, in theory, the maximum electrical efficiency of electrolyzers is close to 120 percent, because an ideal unit would absorb heat from its surroundings and convert this into hydrogen also. An operating efficiency of around 100 percent thus appears to be a reasonable practical goal for commercial installations. The overall efficiency of hydrogen production would then be limited by the efficiency of generation of electricity, which might be expected to be between 35 and 45 percent.

One way to sidestep this limitation is direct thermal decomposition of water within a reactor. This process requires temperatures of about 2500°C, however, and commercial nuclear reactors are not expected to operate at temperatures much above 900°C. But it should be possible to use a multistep reaction that can be carried out at lower temperatures.

Gianfranco de Beni and Cesare Marchetti of the Euratom research center in Ispra, Italy, have developed a four-stage reaction sequence in which dissociation of water is accomplished at temperatures below 730°C:

$$CaBr_2 + 2H_2O \rightarrow Ca(OH)_2 + 2HBr \tag{1}$$

$$Hg + 2HBr \rightarrow HgBr_2 + H_2 \tag{2}$$

$$HgBr_2 + Ca(OH)_2 \rightarrow CaBr_2 + HgO + H_2O \tag{3}$$

$$HgO \rightarrow Hg + \tfrac{1}{2}O_2 \tag{4}$$

The overall reaction is thus

$$H_2O \rightarrow H_2 + \tfrac{1}{2}O_2$$

with a thermal efficiency, de Beni contends, of greater than 50 percent.

The reaction sequence has been tested only in the laboratory, but de Beni hopes to build a pilot plant at Ispra. Many problems remain: The kinetics and equilibria of the four steps are still somewhat uncertain, and materials to contain the corrosive intermediates at high temperatures and pressures will require further development. Nonetheless, it is possible that the Euratom sequence or one like it will eventually be more efficient than the combination of electrical generation and electrolysis.

A third alternative, proposed by Bernard L. Eastlund of the Atomic Energy

Table 14. Annual energy costs per household for various alternative energy systems.

System	Annual cost per household	
	Aboveground electric transmission	Underground electric transmission
Coal-electric and gasoline	$764	$1239
Coal conversion to methane	515	642
Coal conversion to hydrogen	486	613
Hydrogen derived from nuclear energy and coal	815	1290
Nuclear energy conversion to hydrogen	928	1055
Nuclear-electric and gasoline	822	1297
Nuclear—all electric	815	1504

Consumption per household is the same in all cases, and is distributed in the following manner: space heating, 75 million Btu/year; air conditioning, 25 million Btu/year; water heating and cooking, 15 million Btu/year; miscellaneous electric, 25 million Btu/year; automobile, 25 million Btu/year.
[Source: W. E. Winsche, K. C. Hoffman, and F. J. Salzano, Brookhaven National Laboratory]

Commission, would use ultraviolet radiation from the plasma of a fusion reactor for direct photolysis of water vapor. Injection of heavy elements such as aluminum into a hydrogen plasma, he says, would produce photons of the correct wavelength for photolysis, which would then take place in water vapor circulating around the plasma chamber. Experimental confirmation of Eastlund's scheme will require development of an operating fusion reactor, but it seems possible that such a process, in conjunction with electrical generation, could increase the efficiency of a fusion power plant.

The final alternative, proposed by Lester O. Krampitz of Case Western Reserve University, Cleveland, Ohio, is direct production of hydrogen from water by algae. During photosynthesis some algae can increase the oxidation potential of electrons from water to a level as much as 0.3 volt more negative than the hydrogen electrode. It is thus possible, he contends, to couple the reducing potential of these electrons with a hydrogenase enzyme from algae or bacteria and convert hydrogen ions to hydrogen gas. Only preliminary work has been done on the project, but Krampitz recently received a National Science Foundation grant to continue the work under the Research Applied to National Needs program.

Whatever method is used to produce hydrogen, large amounts will be required. Current U.S. consumption of hydrogen is slightly more than 2.28 trillion scf per year, and world consumption is about 6 trillion scf. According to the Institute of Gas Technology, it would require almost 60 trillion scf of hydrogen simply to provide the energy equivalent of U.S. natural gas consumption in 1968. With current electrolyzer efficiencies, production of that amount of hydrogen would require more than 1 million megawatts of electricity, or more than three times the current U.S. generating capacity. Replacement of fossil fuels with hydrogen for all uses other than the generation of electricity, the institute estimates, would require 295 trillion scf by the year 2000.

How much this hydrogen will cost is anybody's guess, but it is certain that its cost will reflect that of electricity. Gregory estimates it would cost $1.50 to $2.50 per million Btu (natural gas currently wholesales for $0.50 to $1.00 per million Btu) if electricity costs 0.4 to 0.7 cent per kilowatt-hour. This estimate may be somewhat optimistic, since current costs of electricity vary from about 0.6 to 0.9 cent per kwhr and many groups expect its cost to rise rather than fall. The cost of hydrogen could thus easily be double Gregory's estimate. The limiting factor, however, will not be the absolute cost of hydrogen, but its cost relative to fossil fuels, and the price of these fuels is certain to climb steeply during the next 30 years.

The availability of relatively inexpensive hydrogen in large quantities could bring about extensive changes in U.S. industry. During the transition to a hydrogen economy, it might be used in the production of other fuels. Coal gasification, for example, is expected to require about 1500 scf of hydrogen for every 1000 scf of methane produced. Most gasification proposals plan to obtain this hydrogen from the coal itself, so the availability of alternative sources would greatly increase the amount of synthetic natural gas that might be produced from a given quantity of coal and extend the lifetime of coal reserves.

Petroleum refining uses increasingly large amounts of hydrogen to improve the quality of the products and to desulfurize the crude oil. By the year 2000, the American Petroleum Institute predicts, a consumption of more than 600 scf of hydrogen per barrel of crude oil will be commonplace. Production of shale oil would require even more processing, and perhaps as much as 1300 scf of hydrogen per barrel.

Inexpensive hydrogen might also make possible the production of synthetic fuels that are not derived from fossil sources. K. R. Williams and N. L. Campagne of Shell International Petroleum Company, London, suggest that it might be possible to make methanol or Fischer-Tropsch gasoline from atmospheric carbon dioxide for as little as 30 to 45 cents per gallon with low-priced hydrogen. Others have suggested that calcium carbonate would be a good carbon source for the production of these fuels.

Ammonia synthesis now accounts for more than 40 percent of hydrogen consumption, and the quantities of hydrogen required can be expected to increase as demand for fertilizer rises. Hydrogen could also find wide use for the direct reduction of iron ore, particularly as supplies of coking coal become more expensive. Production of 1 ton of iron, the American Iron and Steel Institute estimates, would require some 20,000 scf of hydrogen.

A hydrogen economy would also make available unprecedented quantities of cheap oxygen. By weight, oxygen is currently the third largest industrial chemical produced in the United States—11.3 billion kilograms, or about 313 billion scf in 1971—but the quantities produced in a hydrogen economy would dwarf these figures. Credits obtained in the sale of by-product oxygen would reduce the cost of hydrogen, and promotion of new or expanded uses for oxygen might be necessary to make hydrogen production economical. Some of the oxygen could be used to support combustion in applications where atmospheric gases are not desirable, such as in coal gasification. Large quantities would also be used in the production of steel. An even larger market might be developed in the treatment of municipal sewage. For now, however, the use of this oxygen remains the least-studied of all aspects of the hydrogen economy.

V.

ENERGY CONSERVATION

"Energy conservation must become an integral part of future energy policy."

—Senator Henry M. Jackson

19.
ENERGY CONSERVATION

Energy conservation seems destined to become as popular as protection of the environment. Indeed, the two are not unrelated—energy production and use cause much environmental damage. Emissions from cars and power plants foul the air; oil spills, drainage from coal mines, and waste heat pollute bodies of water; uncontrolled strip mining destroys the land itself. Widespread use of nuclear power carries the threat of accidental contamination of the environment with radioactive substances. Some pollution problems are caused by failure to employ environmental safeguards and to develop less polluting technologies, rather than by energy production and use *per se*. But there is no question that conserving energy would lessen the impact on the environment.

The difficulty is that slowing the growing demand for energy as a matter of either public policy or private practice runs contrary to the trend of the last several decades. Consumer decisions to buy a larger car or to install an air conditioner, and business decisions to build poorly insulated houses or to switch from glass bottles to aluminum cans, have been made without any consideration of their implications for energy consumption. An abundance of energy has become synonymous with a high standard of living and a healthy national economy. Indeed, a correlation between the gross national product and total U.S. energy consumption is often cited as evidence that continued economic growth depends on ever greater supplies of energy. The result is a society that uses energy freely, often in inefficient and wasteful ways.

American attitudes toward energy were not always so cavalier. When wood was a primary fuel and had to be cut by hand, it was used sparingly; in the early years of this century, turning off electric lights was a common habit. Nor was energy conservation just a matter of personal values—energy was expensive. But between World War II and 1971, energy prices dropped steadily and there were no major shortages. What might have been seen as a warning sign—the declining overall efficiency of energy use in the United States since 1967—was largely ignored. Now both higher prices and temporary shortages seem inevitable.

The efficiency with which energy is produced plays a major role in determining how much energy is available. The generation of electric power, for example, has increased from about 5 percent efficiency in 1900 to about 33 percent now. But the trend to more efficient generating plants has slowed in recent years, and with the introduction of large numbers of nuclear power plants, average efficiency may even decrease.

The recovery of fuels from the ground also leaves something to be desired. On

Table 15. Distribution of raw energy resources as a percentage of total U.S. energy production in 1970 (6.6×10^{19} joules).

Resource / *Use*	*Indus-trial*	*Trans-porta-tion*	*House-hold and commer-cial*	*Electric power*	*Total*
Oil	7.4	22.9	9.2	3.5	43.0
Natural gas	15.3	1.0	10.7	5.8	32.8
Coal	8.1	0	0.6	11.4	20.1
Hydro and nuclear	—	—	—	4.1	4.1
Total	30.8	23.9	20.5	24.8	100.0

average, only about 31 percent of the oil in domestic reservoirs is recovered, although recovery varies between 5 and 80 percent, depending on the physical properties of the reservoir, the type of oil, and the recovery methods used. Secondary methods such as hydraulic fractioning of the reservoir or flooding with water to increase pressure often improve the recovery, and these methods are widely used. Tertiary recovery methods—injection of chemicals with detergent-like properties to lower the viscosity of the oil, injection of hot steam, or *in situ* combustion of part of the oil (fire-flooding)—are being studied. Advanced recovery methods can be best applied when an entire reservoir is treated as a unit, however, and multiple ownership of reservoirs has led in many instances to overdrilling and other inefficient practices. Oil companies are often not willing to share information on recovery techniques. Yet improving recovery to 60 percent would increase known domestic reserves by 120 billion barrels, a 24 year supply at the current rate of consumption. Even a 10 percent increase in recovery efficiency would at least temporarily eliminate the need for imported oil, although the added cost of recovery could raise oil prices.

Underground coal mining is not much more efficient, but it could be. The room and pillar mining technique used in the United States—in which large parts of a seam are preserved as roof supports—leaves about half the coal in the ground. Longwall mining, developed in European coal fields, permits 80 to 90 percent recovery and greater safety for miners. Strip mining is also about 80 percent efficient, although many of the huge U.S. coal deposits are too deep to permit stripping. Preliminary experiments aimed at extracting energy from deep coal deposits without mining—by means of *in situ* gasification—have begun.

Nearly 80 percent recovery of natural gas is attained from most domestic wells. However, large deposits of gas occur in reservoirs that are too impermeable to be economically drilled unless new recovery methods, such as fracturing with nuclear explosives, can be safely developed. Recovery of gas from coal seams may also be possible with improved mining techniques.

If improving the efficiency of energy production would increase the supply, more efficient use of energy would reduce the amount that is needed. U.S. energy consumption is increasing rapidly at present. Whether this growth will continue at about 3 to 5 percent per year until the end of the century, as some forecasts predict, or will taper off as higher prices and conservation policies have their effect, is hotly debated. Still more controversial is the question of how long unlimited growth can continue in a world with finite resources. Americans use a prodigious amount of energy, more than a third of the world's consumption. Per capita use of energy in the United States is 6 times the average for the rest of the world, so that if other nations' appetites for energy approach the U.S. level, the strain on resources will be truly enormous.

At present, 25 percent of the raw energy resources consumed in the United States are used to generate electricity and this percentage is rapidly increasing. Almost half of the electricity is consumed by industry, and slightly more than half goes to residential and commercial use. Transportation also consumes 25 percent of total energy resources; industry accounts for 30 percent; and residential and commercial uses, largely heating and cooling, consume 20 percent. More efficient use of energy through curtailment of wastes and improved technologies is one way of buying time and stretching out the life of domestic energy resources. Conservation policies that give precedence to the most essential uses of energy are another.

20.
CONSERVATION OF ENERGY
The Potential for More Efficient Use

Five-sixths of the energy used in transportation, two-thirds of the fuel consumed to generate electricity, and nearly one-third of the remaining energy—in all more than 50 percent of the energy consumed in the United States—is discarded as waste heat. Yet little research is devoted to energy utilization. Buildings are designed and constructed with inadequate insulation, large window areas, and excess ventilation—all of which increase the energy needed for heating and cooling. Furnaces and other industrial equipment are often inefficient. Automobile engines are often far more powerful than is necessary. Both more efficient technologies and more rational utilization of energy could save significant amounts of energy.

Conservation of energy will be the more necessary as fuels become scarce and as the environmental problems associated with energy production and use increase. Slowing the rate of growth of energy use through conservation measures would reduce the United States' increasing dependence on imported fuels and would allow more time for the development of improved, less-polluting energy systems. The conservation of energy is therefore a worthy and increasingly important goal. And, despite skepticism on the part of some observers as to the feasibility of wholesale changes in consumer habits and preferences, significant economies appear to be possible, many of which involve little or no change in life-styles. Savings of even 1 percent of the more than 63×10^{15} Btu consumed annually in the United States would represent a significant gain in energy—equivalent to 100 million barrels of petroleum.

Energy consumption in the United States has doubled in the past 20 years and is still increasing rapidly. Use of electricity and natural gas has risen twice as fast. At present, the largest single use of energy is for transportation, but space heating in residences and commercial buildings is also a major use (Table 16). Industrial applications of energy in the form of process steam, direct heat, and electric drive account for significant portions of total energy consumption. Other applications consume only a few percent each, and of these, air conditioning is the most rapidly growing, increasing two and a half times as fast as total energy consumption.

Where and how might more efficient uses of energy be achieved? The largest energy savings and perhaps in the long run the easiest to accomplish could come in homes and commercial buildings, which are seldom designed to conserve energy. Closest to a national standard for insulation in residences are guidelines written by the Federal Housing Administration (FHA). In 1965, these permitted heat losses as high as 50 Btu per square foot of floor space per hour. Revised FHA guidelines issued in 1971 reduced this figure somewhat, but almost none of the

Table 16. Energy consumption in the United States by end use 1960–1968 (trillions of Btu and percent per year).

Sector and end use	Consumption		Annual rate of growth	Percent of national total	
	1960	1968		1960	1968
Residential					
Space heating	4,848	6,675	4.1%	11.3%	11.0%
Water heating	1,159	1,736	5.2	2.7	2.9
Cooking	556	637	1.7	1.3	1.1
Clothes drying	93	208	10.6	0.2	0.3
Refrigeration	369	692	8.2	0.9	1.1
Air conditioning	134	427	15.6	0.3	0.7
Other	809	1,241	5.5	1.9	2.1
Total	7,968	11,616	4.8	18.6	19.2
Commercial					
Space heating	3,111	4,182	3.8	7.2	6.9
Water heating	544	653	2.3	1.3	1.1
Cooking	98	139	4.5	0.2	0.2
Refrigeration	534	670	2.9	1.2	1.1
Air conditioning	576	1,113	8.6	1.3	1.8
Feedstock	734	984	3.7	1.7	1.6
Other	145	1,025	28.0	0.3	1.7
Total	5,742	8,766	5.4	13.2	14.4
Industrial					
Process steam	7,646	10,132	3.6	17.8	16.7
Electric drive	3,170	4,794	5.3	7.4	7.9
Electrolytic processes	486	705	4.8	1.1	1.2
Direct heat	5,550	6,929	2.8	12.9	11.5
Feedstock	1,370	2,202	6.1	3.2	3.6
Other	118	198	6.7	0.3	0.3
Total	18,340	24,960	3.9	42.7	41.2
Transportation					
Fuel	10,873	15,038	4.1	25.2	24.9
Raw materials	141	146	0.4	0.3	0.3
Total	11,014	15,184	4.1	25.5	25.2
National total	43,064	60,526	4.3	100.0%	100.0%

Note: Electric utility consumption has been allocated to each end use.
[Source: Stanford Research Institute.]

buildings in use today meet the new standard (which applies only to new construction), and many older buildings have little or even no insulation.

Even the revised guidelines do not require the economically optimum amount of insulation, according to a study by John Moyers of Oak Ridge National Laboratory (ORNL). From calculations for model houses in three different regions of the country—Atlanta, New York, and Minneapolis—Moyers finds that additional insulation in walls and ceilings, weather stripping, foil insulation in floors, and in some regions, storm windows, can be economically justified. These improvements, in addition to saving the homeowner money, would save an average of 42 percent of the energy used for space heating alone, compared to that used in houses meeting the pre-1971 FHA guidelines.

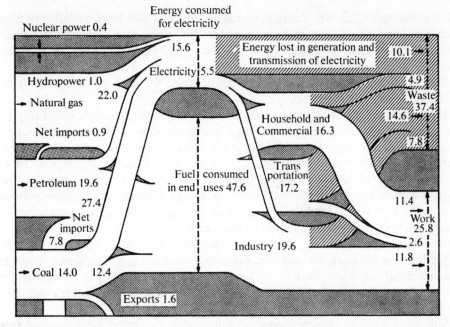

FIGURE 42. Production, consumption, and waste of energy in the United States. Total consumption of energy in 1971 was 63.2×10^{15} British thermal units, excluding nonenergy uses of fossil fuels.
[*Source: Earl Cook, Texas A&M University.*]

Commercial buildings are underinsulated too, according to Charles Berg of the National Bureau of Standards in Gaithersburg, Maryland, so that about 40 percent of the heating energy relative to current practice could be saved in these structures as well. It is difficult to add insulation to existing buildings, but by using adequate amounts in all new construction, energy consumption would gradually reduce over a period of years. The potential savings from both residential and commercial buildings amount to about 7 percent of total national energy use. Actual savings may be even greater, because as insulation is added, the air-conditioning losses are reduced, and the waste heat from lights, stoves, refrigerators, and other appliances becomes a more substantial part of the total heat required.

In addition to structural improvements in the thermal performance of buildings, more efficient heating and air-conditioning equipment is technically possible and in some cases already exists. The efficiency of room air-conditioners sold today, for example, ranges from 4.7 to 12.2 Btu of cooling per watt-hour of electricity. Efficiencies are not ordinarily stated explicitly on commercial units, but manufacturers have been induced to include this information on units sold in the New York City area. Although the amount of energy used for air-conditioning is considerably less than for space heating, it is an important drain on the power supply during peak summer months. Moyers and Eric Hirst, also of ORNL, estimate that upgrading the average efficiency of window units from 6 to 10 Btu per whr (in 1970) would have saved 15.8 billion kilowatt-hours of energy per year.

Furnaces for space heating are typically about 75 percent efficient as sold, losing a quarter of the heat in the exhaust. According to Berg, however, the frequent operation of such furnaces at low capacity and the infrequent maintenance that is common for residential units may lead to efficiencies as low as 35 to 50 percent in practice. Improved design and regular maintenance could substantially improve their performance. Electric resistance heating, which is now being installed in about one-third of all new homes and perhaps half of all new office buildings, is virtually 100 percent efficient in place. However, the production of 1 kwhr of electricity requires on average 3 kwhr equivalents of heat, and since about 10 percent of the electricity is lost in transmission and distribution, the real efficiency of electric heating is less than 30 percent.

Electrically driven heat pumps, which are not now widely used, could improve the efficiency of electric heating because a heat pump delivers on average about two units of thermal energy for each unit of electrical power it consumes; actual performance depends considerably on the climatic conditions of the region. In the past, frequent failures and high maintenance costs have been common, but improved models are now available. Heat pumps may become an increasingly attractive method of space conditioning as fossil fuels grow scarce and nuclear power plants become the prevailing source of electricity.

Solar heating and cooling, although commercially unproved, would be still more attractive as a conservation measure, because it does not consume unrenewable resources. Solar hot water heaters, which it is estimated could replace about half of the conventional water heaters or more than 1 percent of total energy use, have been in limited commercial service for some time.

Still other ways to reduce energy consumption in the home have emerged from an ongoing study directed by D. G. Harvey, of Hittman Associates in Columbia, Maryland, for the U.S. Department of Housing and Urban Development and the National Science Foundation. Harvey finds that decorative outdoor gas lights and pilot lights in gas appliances use surprisingly large amounts of fuel, and he points out that electronic igniters, which could eliminate the need for pilot lights, are commercially available. Outside air flowing in through leaky window frames and open chimney flues in unused fireplaces are also major causes of energy loss. Heat recovery systems, Harvey believes, could reduce flue losses in the heating system and improve gas furnace efficiencies by as much as 12 percent.

Frost-free refrigerators and freezers use about one and a half times the energy of manual defrosting units. Fluorescent lights use a quarter as much electricity as incandescent bulbs. Well-insulated ovens—including most self-cleaning ovens—consume significantly less energy than their poorly insulated counterparts. Less air-conditioning would be needed if small fans were used to ventilate attics. And finally, a deciduous tree near a house can make a noticeable difference, Harvey believes, shielding the roof in summer but letting the sun warm the house in winter.

The rising sales of mobile homes, which now account for one out of every four new dwellings in the United States, may have an increasingly important influence on residential energy use. Because of their thin walls, limited insulation, and box-like construction, mobile homes are high users of energy, often requiring inefficient space heaters in winter and several window air-conditioning units in summer. Existing standards for these relatively inexpensive, factory-built homes were not written with energy conservation in mind.

Architectural practices often promote excess energy use, according to Richard Stein, of Richard G. Stein and Associates in New York City. He points out that poor design often results in the overuse of steel, concrete, and other energy intensive building materials by as much as 50 percent. The illumination levels recommended in commercial buildings have more than tripled in the last 15 years and there is now considerable disagreement as to whether such high illumination—100 foot candles in many office applications—or uniform intensity of lighting is necessary or desirable. Nearly a quarter of all electricity is used for lighting, and Stein believes that a 4 percent savings in total electricity use could be achieved immediately by reducing excess lighting in existing buildings and by more effective use of lighting in new buildings.

Modern high-rise office buildings consume inordinate amounts of energy. The World Trade Center, for example, a 110-story structure in New York City, uses 80,000 kw, more than the entire city of Schenectady, New York (population, 100,000), and the trend to such buildings is accelerating in urban areas. Stein believes that electrical heating is particularly inefficient in such buildings, because they can readily use fossil fuels instead. Other energy savings are possible with reflective window glass, reduced air ventilation, and absorption central air-conditioners that operate on heat, not electricity. In all, Stein believes that careful design might reduce the energy needed to operate high-rise buildings by half. Reductions in peak power demand would be even greater.

With some exceptions, there is evidence that efficiency of energy use has not been a subject of concern in many industries, although industry consumes the largest share of U.S. energy production. The production of primary metals, basic chemicals, petroleum products, food, paper, glass, and concrete account for most of the energy used. As fuel prices rise, industry can be expected to institute substantial economies. The energy required to produce a ton of steel, for example, declined by 13 percent to 26 million Btu between 1960 and 1968—primarily because of more efficient blast furnaces—and is expected to decline still further. New vacuum furnaces developed for industrial use require only a quarter of the energy consumed by earlier designs. Heat recovery devices and better thermal management of many processes may also save considerable amounts of energy. On the basis of these examples and others, Berg estimates that as much as 30 percent of the energy used by industry might be conserved. And because corporate management can respond more rapidly to changing conditions than the individual consumer or even than the fragmented construction industry, many observers believe conservation measures in industry have the greatest potential for short-range impact on the demand for energy.

A case of particular interest is the utility industry, which has improved the efficiency of electrical generation from about 5 percent in 1900 to nearly 40 percent in the newest coal-fired plants. Oil- and gas-fired plants are slightly less efficient, and the average for all existing power plants is about 33 percent. Nuclear power plants with light water reactors convert only 32 percent of the heat they generate into electricity; however, high temperature gas reactors that are now becoming available have efficiencies of nearly 40 percent. The development of combined cycle power plants—with high-temperature gas turbine or magnetohydrodynamic generators in addition to steam turbines—could increase generating efficiencies to 50 or 60 percent. At present, however, the generation, storage, and distribution of electric energy is inherently inefficient. As long as fossil fuels are consumed to generate substantial amounts of electricity, a state of affairs that is expected to prevail throughout this century, the use of electric power for applications where the direct combustion of fossil fuels will do is clearly wasteful.

Transportation constitutes the largest single end use of energy, but opportunities for significant saving appear to be less, because changes to more efficient modes of travel involve changes in life-styles that are more substantial than the changes necessary for most of the conservation measures discussed previously. A study by Hirst at Oak Ridge reveals that during the 1960's passenger traffic on U.S. rail-

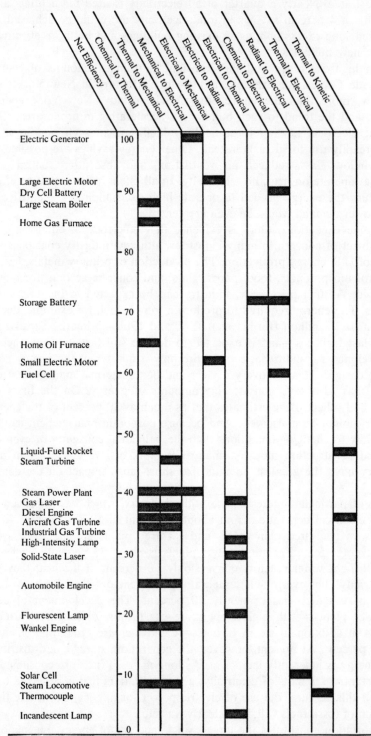

FIGURE 43. Efficiency of energy converters runs from less than 5 percent for the ordinary incandescent lamp to 99 percent for large electric generators. The efficiencies shown are approximately the best values attainable with present technology.

[*Source: From* Conservation of Energy, *Committee on Interior and Insular Affairs, United States Senate, 1972.*]

Table 17. Energy efficiency for passenger transport.

Item	Btu per passenger mile	
	Urban	Intercity
Bicycle	200	
Walking	300	
Buses	3700	1600
Railroads		2900
Automobiles	8100	3400
Airplanes		8400

[Source: E. Hirst, Oak Ridge National Laboratory]

Table 18. Energy efficiency for freight transport.

Item	Btu per ton mile
Pipeline	450
Railroad	670
Waterway	680
Truck	3,800
Airplane	42,000

[Source: E. Hirst, Oak Ridge National Laboratory]

roads decreased by half, automobile mileage increased by 50 percent, and airline mileage increased nearly threefold. Yet the energy efficiency is higher for railroads than for cars and airplanes (Table 17). Freight transport also shows shifts from railroads to trucks and airplanes, more energy-intensive modes of transport (Table 18).

Clearly most people prefer to go by car, despite the relatively high energy cost associated with this vehicle. Indeed, Hirst estimates that when both the direct and indirect energy costs are included, the automobile accounts for about 21 percent of total U.S. energy consumption. Yet the standard American car gets only 12 miles per gallon of gasoline, roughly half that of most European cars. In part, the decline of rail service and urban mass transit relative to automobiles and airplanes reflects the greater convenience, flexibility, and speed. But governmental promotion of automobile, truck, and airplane traffic through the subsidy of roads and airports has also undoubtedly been influential. Reversing these shifts in passenger and freight traffic could save significant amounts of energy, predominantly in the form of petroleum—thus reducing the need to import this commodity.

Still other methods of conserving energy include central heating plants for groups of buildings, and in some cases whole towns. These have been used occasionally both in the United States and abroad and can make efficient use of waste steam from a power plant. Total energy systems in which small gas turbines or fuel cells generate electricity locally, in addition to providing heat, could substantially increase the overall efficiency of energy use, although the operation and maintenance of such facilities pose difficult and costly problems.

Although it seems unlikely that even extreme conservation measures can entirely halt the need for more energy, it is undeniably a poor idea to perpetuate wasteful and often ineffective uses of energy. The potential for reducing the demand for energy by means of more efficient use of energy resources appears to be enormous, amounting ultimately to perhaps 25 percent of what would otherwise be consumed. Under present practices, energy that could otherwise be saved is wasted in buildings, in industrial processes, and in transportation. Available techniques for utilization of waste heat, for production of more efficient machinery, and for prevention of energy losses are seldom applied, in part because the cost of energy has been low. To bring about some of these potential savings, financial incentives and other means of changing attitudes and habits in energy use will be necessary. Far from being an unrealistic notion, conservation is clearly a major energy option.

21.
ENERGY NEEDS
Projected Demands and
How to Reduce Them

For the last two decades, energy has been an inexpensive and readily available commodity in the United States. Indeed, the cost of energy has declined relative to overall price indices in the period since World War II. But there is evidence that a new era of energy use is beginning, ushered in by growing shortages of the traditional fuels and characterized by rising prices for energy. During 1971 for example, the average price of electricity (measured in constant dollars) increased for the first time since 1946. Some observers have speculated that, by the end of the century, the cost of electricity might double, the cost of gas (natural and synthetic) might triple, and the costs of petroleum, coal, and uranium might increase substantially.

The demand for energy is sometimes considered to be independent of its price, and many projections of future energy needs are made with that assumption. Many economists, however, believe that higher prices will slow the growth in the use of energy and, because of energy's central role, may even retard the growth of the economy as a whole. The desirability of slowing economic growth is still controversial, but damping the demand for energy is beginning to receive serious consideration in the federal and state governments for other reasons. Environmental problems caused by energy production and use, supply problems caused by shrinking domestic reserves of gas and oil, and the economic penalties of importing large amounts of fuel could all be alleviated by cutting energy demands. Higher prices, tax incentives and subsidies (or their removal), changes in building codes, restriction of advertising that encourages energy use, and even rationing have been proposed as measures toward that goal.

The demand for electricity has grown more rapidly than that for other forms of energy in the last two decades, and the need for electricity (and for more power plants) has occasioned considerable debate. The Federal Power Commission (FPC) estimates that the demand will double from about 1.5 trillion kilowatt-hours in 1970 to 3 trillion in 1980 and will almost double again by 1990; similar estimates are that demand will continue to grow rapidly until the end of the century. The FPC estimate and others like it are based at least implicitly on extrapolations of previous trends in overall economic and population growth, and not on any detailed model of the manner in which these and other variables that affect the demand for power might change. Hence these estimates are likely to be accurate only to the extent that past trends continue essentially unchanged into the future.

More sophisticated studies of the demand for electricity under a variety of alternative assumptions about the future are beginning to appear. In a 2-year study

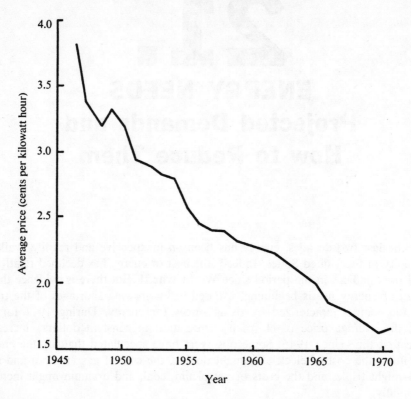

FIGURE 44. Ratio of the average price of electricity to the gross national product inflation index. The average price of electricity declined relative to other prices from 1946 to 1970, but increased in 1971.
[*Source: From "Electricity Demand Growth and the Energy Crisis," by D. Chapman, T. Tyrrell, and T. Mount. Science, November 1972.*]

funded by the National Science Foundation's program of Research Applied to National Needs, for example, D. Chapman and T. Mount of Cornell University and T. Tyrrell of Oak Ridge National Laboratory evaluated the relations between variables that might influence the growth of demand and the actual demand for electricity, state by state, for each class of consumers—residential, commercial, and industrial. Their econometric analysis indicates that the price of electricity is the most important determinant of growth of electricity use for all types of consumers, followed by population growth, personal income, and the price of natural gas. Thus substantial cost increases and reduced population growth, according to the Cornell–Oak Ridge study, would significantly reduce the demand for power in the long run.

The downward trend in the relative price of electricity has apparently begun to reverse itself, and prices may well rise sharply as fuel scarcities, the rising cost of power plants, and pressure to incorporate the social and environmental cost of energy production in its price tag make their impact felt (Fig. 44). The population growth rate appears to be declining because of decreases in both the birth rate and the fertility rate. By the end of 1972, fertility had fallen to about 2.08 children per woman, below the "zero population growth" replacement rate. Whether these trends will continue is uncertain, predictions about the future being what they are, but it is possible to calculate the demand for power that would result if they did. Chapman and his associates, using their econometric model, find that variations in the price of electricity and to a lesser extent in the rate of population growth can cause as much as a fivefold reduction in the electricity needs projected by the FPC (Fig. 45).

If the relative cost of electricity doubles by the end of the century, the model projects a demand for about 2 trillion kilowatt-hours, only 33 percent more than that used in 1970. If the cost of electricity increases only slightly (in constant dollars)—which the FPC believes it will—Chapman finds a demand of about 3.5 trillion kwhr, much less than FPC estimates. Constant prices for the next 30 years would increase the demand still further, according to the model, but a 50 percent decrease in the cost of power would be necessary to maintain a doubling of the demand every 10 years. Hence Chapman believes that the FPC projections are seriously in error and that plans based on those projections will be unrealistic.

Similar evidence comes from a detailed case study of the demand for electricity

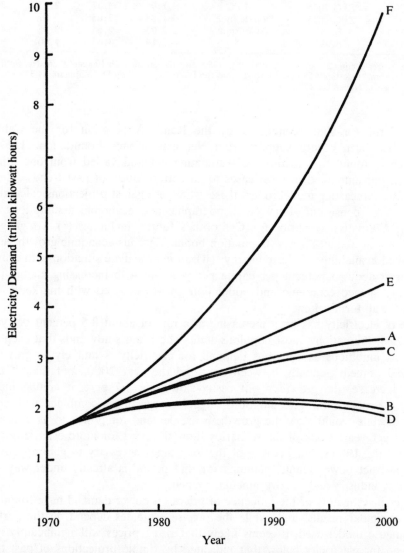

FIGURE 45. Electricity demand projections from Table 19. All projections give similar results for the near future but differ greatly over longer periods of time.
[*Source: From "Electricity Demand Growth and the Energy Crisis," by D. Chapman, T. Tyrrell, and T. Mount.* Science, *November 1972.*]

Table 19. Electricity demand growth and alternative assumptions. BEA, Bureau of Economic Analysis; FPC, Federal Power Commission; ZPG 2035, zero population growth reached in 2035. In the constant price assumption 1970 prices are maintained in each region. In the "double by 2000" assumption, the average price in each region increases annually by 3.33 percent of its 1970 value for 30 years. In case F, the FPC demand projection and the BEA population projections were used. A total of 1.53 Tkwh of electricity was generated in 1970.

			Electricity demand (Tkwh)			
Case	Population assumption	Electricity price assumption	1975	1980	1990	2000
A	BEA	FPC	1.98	2.38	3.01	3.45
B	BEA	Double by 2000	1.88	2.07	2.11	2.01
C	ZPG 2035	FPC	1.98	2.37	2.95	3.29
D	ZPG 2035	Double by 2000	1.88	2.05	2.07	1.91
E	BEA	Constant	2.02	2.54	3.56	4.56
F	BEA	*	2.14	3.05	5.66	9.89

*Average prices decline 24 percent from 1970 to 1980, and 12 percent each 10 years thereafter until 2000. [Source: From "Electricity Demand Growth and the Energy Crisis," by D. Chapman, T. Tyrrell, and T. Mount. *Science*, November 1972.]

in California, a study conducted by the Rand Corporation for the California Resources Agency with support from National Science Foundation. The Rand researchers found that methods of estimating demand varied from one utility to another, but amounted in most cases to an extrapolation of past trends. They developed a forecasting model to test these estimates against projections calculated on the basis of consistent statewide demographic and economic assumptions. They present alternative scenarios for California's future, including (i) a high growth situation that assumes a new population boom, vigorous economic growth, and the continued availability of cheap energy; (ii) an intermediate situation with economic growth at about 3 percent per capita per year but with increasing energy prices; and (iii) slowed economic and population growth coupled with markedly higher prices for all forms of energy.

Use of electricity has been increasing at a rate of about 8.5 percent per year in California, the nation's most populous state. The Rand study finds that even under the most ambitious assumptions demand for electricity is unlikely to grow faster than 6.3 percent annually between now and the year 2000, and that, if energy prices increase, the growth might easily slow to a 4.7 percent annual increase. Under the most limiting assumptions, higher energy prices combined with an economic decline would slow the growth in the demand for power to as little as 3.4 percent per year. None of the scenarios show the need for more electricity changing until the 1980's, but because of the long times necessary to gain approval for and construct power plants, planning for that period is already under way within the utility industry and in governmental agencies.

Other investigators of the influence of prices on power demand have found qualitatively similar results, and, while the precise estimates depend on the particular econometric model used, it seems likely that rising prices will significantly reduce the need for electricity below that indicated by simple projections of past trends. Slower growth rates could alter the prospects for new methods of generating electricity, decrease the urgency with which near-term options—such as the breeder reactor—need to be pursued and buy more time to examine other sources of power.

Slower growth rates for all forms of energy consumption might mitigate the environmental damage associated with energy production and use and ease the shortages of gas and oil that are now expected. Studies of the demand for gas and other fossil fuels which are comparable to the econometric studies of electricity demand are lacking; but Chapman believes that for gas, at least, rising prices will have similar effects. Significantly, both the Cornell–Oak Ridge group and the Rand group found little evidence that gas prices influence the demand for electricity despite the fact that these are often competing sources of energy, implying that rising prices for one would not cause a major switch to the other.

If higher prices could slow the growth in the demand for energy and particularly for electricity, deliberate conservation policies should reduce demand still further. Indeed, the econometric studies suggest that raising prices to reduce demand may be an exceedingly effective means of promoting the conservation of energy.

One method of reducing the demand for power and promoting its efficient use is to do away with rate structures that provide incentives for high volume consumers. In most states, high volume consumers receive such large discounts on the price of electricity that small users effectively subsidize large users. Because the industries and commercial interests that benefit from this discount would be the first to reduce their use of power when faced with higher prices, the method appears to many observers to be a particularly appropriate one. An analysis, conducted by the Environmental Defense Fund (EDF), of the price-sensitivity of the demand for power in Wisconsin led the EDF to intervene before the state utility commission in hearings on proposed rate increases for electricity. The EDF claims that by pricing power in proportion to the true cost of supplying each customer, the demand for electricity in Wisconsin would be sufficiently decreased to obviate the need for several projected power plants. The environmental group plans to challenge electricity rates in other states on similar grounds.

Others have proposed more drastic changes in the cost of energy and in the tax system as a means of slowing the exploitation of energy resources. Walter Heller of the University of Minnesota, for example, believes that large depletion allowances, capital gains shelters, and special tax deductions should no longer be allowed for energy-producing industries: "Here is another case where the believers in the market-pricing system ought to live by it. The public is subsidizing these industries at least twice—once by rich tax bounties and once by cost-free or below cost discharge of waste and heat." The principle put forward by Heller and other critics of existing financial incentives is that the cost of energy should reflect the environmental and ultimately the social costs of producing it. Policies such as these, whatever their other effects, would result in higher prices for energy and hence greater incentives for its conservation.

In addition to tax policies designed to raise prices for traditional sources of energy, tax incentives have been proposed to encourage more efficient building design, more efficient appliances, and the use of solar energy. Energy used in buildings and appliances could also be reduced by regulatory policies. Federal guidelines and state building codes, for example, could be changed to require more insulation in houses and heat-reflecting glass in office buildings. Requiring appliances to bear labels that make explicit the efficiency of the device and the estimated operating costs would allow consumers to make more informed decisions. Many of these proposed policies would not discourage economic growth—a consequence feared by many who oppose efforts to reduce the demand for energy; others, such as requiring electric utilities to deduct the cost of promotional advertising from profits rather than from taxable income (as is now done), might have economic side effects.

How feasible are such policies and how much might they reduce the demand for energy? Very little research has been directed to these questions, but a second Rand study, conducted for the California state legislature, claims that such measures would be very effective. By the year 2000, use of electricity in the state might be reduced by as much as 430 billion kwhr annually through conservation policies, a 50 percent reduction in the demand projected by conventional methods. These savings would reduce the need for new power plants from an estimated 127 to 45 or less. Environmental damage due to power generation would also be reduced, the study concludes, but only relatively minor economic dislocations would occur and the growth of the state's economy would not be affected. This finding has led some federal officials to question whether increasing energy use and economic growth are necessarily correlated, as is often assumed. The Rand study recommends an early introduction of conservation measures.

Research on ways to improve the efficiency of energy use and on means to implement these improvements is just beginning, and more is needed. But it seems likely that from now on energy will always cost more and that expensive energy will induce some consumers to do with less and others to use it more efficiently. As a result, energy needs in this country may well be grossly overestimated. Moreover, it seems clear that in the long run energy needs could be reduced still further through effective conservation policies.

VI.
ENERGY POLICY

"We have a National Energy Policy and it's basically one of promotion, of pushing the product."

—S. DAVID FREEMAN,
Director, Ford Foundation Energy Policy Project

22.
ENERGY AND THE FUTURE
Research Priorities and Energy Policy

The energy problems facing the United States are only partially amenable to technological solutions. Not every new energy conversion device nor every exotic energy source needs to be developed. More significant, in the long run, will be new attitudes and policies that take into account finite resources and equitable distribution of the costs of producing energy. But it is certain that better methods of extracting energy from coal, for example, and more efficient means of using energy will be needed well before the end of the century. Heat and electricity from solar and geothermal sources may also be essential by then, as may the improved utilization of uranium resources in breeder reactors. But these and other technologies will not be available unless more deliberate, well-funded and well-managed efforts to develop them are undertaken.

How soon the technology to exploit new energy sources or to improve the use of existing sources will be available depends partly on how much money and effort are devoted to their development. Hence the extent to which technology could help to solve the larger energy problems—the conflict between environmental standards and energy needs, the massive waste of energy by inefficient technologies and careless practices, the balance of payments deficits and national security worries arising from projected huge imports of oil—will depend on the priorities for energy R & D. The diverse groups clamoring for a national energy policy do not agree on what these priorities should be. There is, however, general agreement among environmentalists, industrialists, university scientists, and government officials that the present distribution of research efforts is greatly imbalanced and that much more money for research should be spent both in industry and government. A task force established by the utility industry, for example, recommended research funding of more than $1 billion per year for electrical energy alone, an amount nearly double existing outlays.

This book has focused on specific technologies for supplying energy and on the "technology" of energy conservation. The state of knowledge concerning many of these technologies is rudimentary and few are free from potential environmental problems and substantial engineering difficulties. At the same time many clearly have high potential for contributing to the resolution of current and prospective energy problems, and many deserve far more serious investigation. Present R & D patterns—largely the result of historical precedents, such as the wartime development of nuclear energy by the government and past legislative biases toward oil and traditional funding mechanisms are not adequate to meet the need In discussing R & D priorities, a useful starting point is to consider the nature of the energy problems that confront the United States and to sample some differing views on how they might be faced.

Energy prospects in the United States are often classified as near term (between now and 1985), intermediate (1985 to the end of the century), and long range. Problems of the immediate future may be the most difficult problems to resolve, because of the time required to develop new technology and implement new policies. Chief among these will be the problem of providing enough electricity with a coal-based utility industry while meeting environmental constraints, either through improving coal technology or converting to other fuels, predominantly oil. Nuclear power is still in its infancy (less energy was obtained in 1972 from uranium than from burning wood) and, even in forecasts by its more optimistic proponents, cannot expand rapidly enough to replace fossil fuels as the main source of electricity much before the year 2000.

Conversion to clean fuels in power plants will aggravate a second and ultimately more intractable supply problem, the growing shortage of domestic oil and natural gas and the increasing reliance on imported fuels in ever greater quantities. The trend to all-electric homes and office buildings will also intensify fuel shortages as long as oil and natural gas are used to fuel a substantial number of the power plants, as they do increasingly in urban areas. But even if sulfur emissions from coal-fired generating plants can be reduced, oil and natural gas will remain in great demand in industry because they are so much easier to use. In the absence of urban mass-transit systems, the growing transportation system will continue to depend almost entirely on petroleum fuels.

Oil is the main source of energy for the United States and, in the absence of deliberate policies to the contrary, is likely to remain so for the rest of the century. Imports of oil, according to projections by the National Petroleum Council (NPC), will rise to 19 million barrels per day by 1985 if present trends continue. This quantity would represent a third of the nation's total energy supply and nearly two-thirds of its petroleum supply, and would cost an estimated $32 billion per year. Draining capital from the country at such a rate might have ruinous economic effects. More significantly, the extent of projected dependence on oil from the Middle East, the availability of which could not be assured, might pose a serious threat to national security.

The intermediate period will undoubtedly be characterized by continuing international competition for oil and gas and the marked depletion of domestic supplies. Conversion of coal to synthetic fuels may become a major industry, and the extensive development of new energy sources such as oil shale, geothermal energy, and solar energy may begin in earnest. Breeder reactors may approach commercial readiness, expanding the potential of nuclear power as a source of electricity. Efforts to use energy more efficiently and possibly to reduce the demand for energy are likely to become more significant. Energy will in all probability cost much more than it does today.

For the long run there is no shortage of potential energy sources. Either fusion or solar energy and probably both might eventually become the prime sources of energy provided that the necessary technologies can be developed and made economically competitive—a proviso with no certain timetable. Geothermal energy, nuclear fission, and coal might also contribute significantly for centuries, although the potential social and environmental hazards associated with fission may restrict its use and coal deposits may become more valuable as a source of chemicals than as fuel. Energy will be consumed predominately in the form of electricity and synthetic fuels, possibly hydrogen in many applications.

Difficulties associated with the supply of energy are not the only problems that must be resolved. Among those frequently mentioned in discussions of national policy are:

► Reconciling environmental policies with energy policies. For example, should the oil industry be allowed to drill for offshore deposits along the Atlantic coast, where much of the nation's undiscovered reserves are expected to be; or should preservation of marine fauna and the esthetic and recreational value of coastal areas take precedence? Are the reduced exhaust emissions from automobiles to be required in coming years worth the 30 percent increase in fuel consumption that will result?

► Balancing foreign and domestic supplies. The oil industry claims that low-cost foreign oil should be kept out of the United States for national security reasons —a policy some have labeled "drain America first." Others feel it should be imported while it is cheap to hold down energy prices, and that provisions for security, such as storage of a substantial reserve, could be made. Still others believe that importation of oil will ultimately force the United States to compete efficiently or lose its great power status.

► Energy prices. The gas industry would like natural gas prices to be removed from regulatory controls and allowed to rise sharply in order to promote exploration and reduce demand; but such a policy would bring windfall profits to industry and the burden of higher energy prices to the poor. Unresolved questions are whether higher prices would actually lead to finding new supplies and whether gas and electricity rates should continue to promote the use of these energy forms by discounts to large consumers.

► Energy monopolies. Are the traditional incentives for fair prices, efficient management and substantial research efforts in energy—largely competition between different fuels—still possible? The trend in recent years is for oil companies to buy up or merge with coal companies, to acquire uranium reserves and other segments of the nuclear industry, and to develop geothermal resources. Does this mean greater governmental control over energy matters is needed?

Differing views have been advanced as to what the nation's energy policy should be. The American Petroleum Institute, for example, favors the deregulation of gas prices, the rapid leasing of federal lands and offshore areas, and continued restriction of oil imports. In the industry's view, the highest priority must be given to assuring energy supplies, even if this means higher prices and fewer environmental restrictions. In contrast, some environmentalists believe that excessive promotion of energy consumption is the primary problem. They feel present growth rates in the production of energy cannot and should not be sustained.

Many of the energy problems mentioned above and the differing approaches toward their solution are involved in setting research priorities. The allocation of the $600 million that the federal government is spending on energy R & D this fiscal year ($350 million on nuclear fission, $135 million on fossil fuels, $65 million on fusion, and the remainder in small amounts on a host of other technologies) still reflects to a large extent past attitudes on energy priorities.

In retrospect, it seems that the neglect of research on coal technologies in the last 20 years is a major cause of present fuel and electricity shortages. There are ample domestic supplies of coal, but the methods for its recovery, combustion, and conversion to other fuels are still primitive. No effective methods for removing sulfur from stack gas have been found, and the removal of it and other pollutants during or prior to combustion seems likely to be more successful. One possibility, fluidized-bed combustion chambers, has received almost no attention in this country. The conversion of coal to synthetic high-Btu gas for pipeline use and to low-Btu power gas for on-site industrial and utility use is receiving increased support, but far from enough. Without extensive pilot plant trials, there is no way of knowing which of the many possible processes for gasification and desulfurization will turn out to be most advantageous. But these trials could take decades with present funding and contractual arrangements for research. If massive imports of oil are the major economic burden and national security threat that many believe, then coal research should have the highest priority for both the near and intermediate term.

A near-term possibility that could temporarily relieve shortages of clean fuel is the gasification of naphtha and residual oils. The production of low-Btu power gas in particular, which appears to be easier, would allow high-sulfur oils that are readily available from Venezuela to be used in conjunction with combined-cycle turbines for power generation in urban areas. A major problem appears to be the reluctance of the oil industry, where expertise with such processes is to be found, to undertake the necessary development.

Ultimately the direct combustion of coal in magnetohydrodynamic (MHD) generators, the conversion of coal into oil, and the development of oil-shale deposits may prove attractive. But existing small-scale efforts to develop MHD are unlikely to get very far; it seems to be a characteristic of research on this and other energy conversion processes that meaningful progress can be made only when nearly full-scale equipment—large pilot plants and demonstration plants—is tested under realistic conditions.

The development of nuclear power is perhaps the only successful federally sponsored energy research effort and the only one where adequate funding is now available. The development of the breeder reactor has the highest priority in the Nixon energy program, although the technology is primarily a long-range energy source that is unlikely to contribute significantly to U.S. energy production soon. It is nonetheless a major option for the long run. The breeder program has been criticized, however, for poor management and for relying exclusively on one breeder concept. It is no secret that many observers, including federal officials in the now defunct Office of Science and Technology, utility executives, and scientists within the Atomic Energy Commission's own laboratories, believe that a backup concept should also be vigorously developed.

Both nuclear and fossil fuels have existing organizations to fund research and well-established industries to press for their continued utilization. Many of the more unconventional energy sources are at a disadvantage in this respect, and the problems of bringing them into actual use may in some instances—for example, solar heating in homes—be much more difficult than developing the technology.

Of the long-term sources, fusion appears to be the most adequately funded. It is still an open contest as to whether laser-induced fusion or schemes that depend on magnetic confinement of the plasma will ultimately prove more feasible; both should be pursued. Solar energy is much closer to being practical. It is largely the economic competitiveness of solar-thermal and photovoltaic power systems that is in doubt, not—as with fusion—their scientific feasibility. Indeed, development of these systems is considerably less demanding technically than, for example, the breeder reactor. Developing solar energy primarily involves advances in engineering capability and manufacturing techniques—problems that yield readily to large-scale industrial efforts. In addition to its long-range potential, the near-term use of solar energy for space heating and cooling not only appears likely to become economically and technically feasible, but could significantly alleviate shortages of natural gas and electricity. No major development effort for solar energy is now in sight, however, and even preliminary studies are underfunded. A much higher priority, probably greater than that awarded to fusion research, seems warranted.

The prospect of geothermal leases on federal land has stimulated industrial interest in this resource, and exploration techniques, now in an early stage of development, may be expected to improve rapidly. More research on the low-temperature turbines needed to tap these deposits and on reservoir management techniques would facilitate the development of geothermal power plants. In some areas, these could significantly supplement traditional sources of power.

Changing the way energy is used is a greatly underresearched area. Eliminating waste with more efficient technology and energy-saving designs could reduce energy requirements by amounts comparable to projected oil imports. Accordingly, research on energy conservation deserves a high priority. But many conservation measures, such as improved insulation in buildings, will take a long time to achieve full impact. A firm government policy in favor of energy conservation, including the elimination of existing policies that implicitly promote energy use and the investigation of additional measures to slow down the growth rate of energy consumption, would help even more, both immediately and in the long run.

There need be no shortage of energy. But wiser use of what is available now and more effective efforts to provide a range of energy options for the future are necessary.

APPENDIX

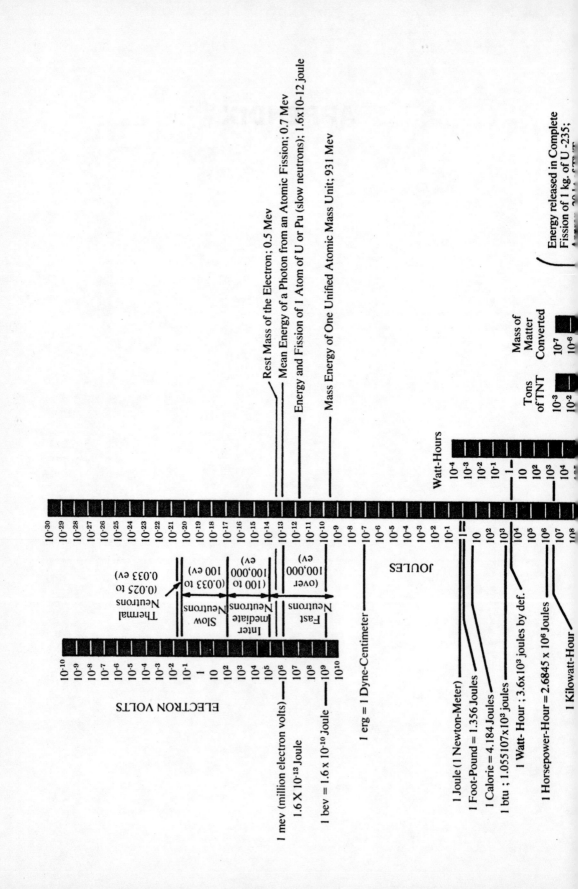

Rest Mass of the Electron; 0.5 Mev

Mean Energy of a Photon from an Atomic Fission; 0.7 Mev

Energy and Fission of 1 Atom of U or Pu (slow neutrons); 1.6x10-12 joule

Mass Energy of One Unified Atomic Mass Unit; 931 Mev

Energy released in Complete Fission of 1 kg. of U -235;

Mass of Matter Converted

10^7
10^6

Tons of TNT

10^{-3}
10^{-2}

Watt-Hours

10^{-4}
10^{-3}
10^{-2}
10^{-1}
1
10
10^2
10^3
10^4

JOULES

10^{-30}
10^{-29}
10^{-28}
10^{-27}
10^{-26}
10^{-25}
10^{-24}
10^{-23}
10^{-22}
10^{-21}
10^{-20}
10^{-19}
10^{-18}
10^{-17}
10^{-16}
10^{-15}
10^{-14}
10^{-13}
10^{-12}
10^{-11}
10^{-10}
10^{-9}
10^{-8}
10^{-7}
10^{-6}
10^{-5}
10^{-4}
10^{-3}
10^{-2}
10^{-1}
1
10^2
10^3
10^4
10^5
10^6
10^7
10^8

ELECTRON VOLTS

10^{-10}
10^{-9}
10^{-8}
10^{-7}
10^{-6}
10^{-5}
10^{-4}
10^{-3}
10^{-2}
10^{-1}
1
10
10^2
10^3
10^4
10^5
10^6
10^7
10^8
10^9
10^{10}

Thermal Neutrons (0.025 to 0.033 ev)

Slow Neutrons (0.033 to 100 ev)

Intermediate Neutrons (100 to 100,000 ev)

Fast Neutrons (over 100,000 ev)

1 mev (million electron volts)
1.6 X 10^{-13} Joule

1 bev = 1.6 x 10^{-10} Joule

1 erg = 1 Dyne-Centimeter

1 Joule (1 Newton-Meter)

1 Foot-Pound = 1.356 Joules

1 Calorie = 4.184 Joules

1 btu ; 1.055107x10^3 joules

1 Watt- Hour ; 3.6x10^3 joules by def.

1 Horsepower-Hour = 2.6845 x 10^6 Joules

1 Kilowatt-Hour

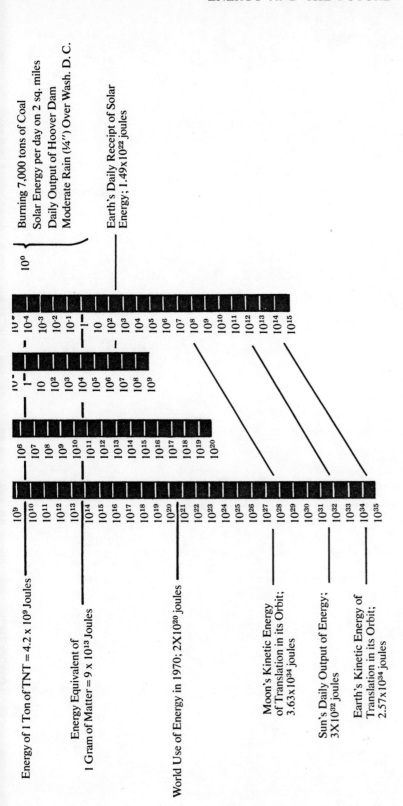

ENERGY UNITS AND CONVERSION FACTORS

This chart shows the energy levels of various real things, from the thermal energy of a neutron at 10^{-20} joule to the kinetic energy of translation of the earth in its orbit at 10^{34} joules. Energy increases with descent down the scales, and is constant along a horizontal line. The energy equivalent to conversion of one gram of matter, for example, is 9×10^{13} joules, which is approximately 2.5×10^{10} watt-hours, or the energy derived from the explosion of 20 kilotons of TNT, or the energy released in the complete fission of one kilogram of uranium-235.

[*Source: From "The New College Physics: A Spiral Approach," by Albert V. Baez. Copyright © 1967 by W. H. Freeman & Company.*]

GLOSSARY

On the following pages are defined some of the terms encountered most commonly in discussing energy, both in this book and in general usage. The list is by no means exhaustive, but it should provide a brief introduction to most of the key concepts in energy technology.

Italicized words within a definition are defined elsewhere in the glossary.

AGGLOMERATION The aggregation of small particles of a solid into larger clumps. Many *coals* exude a sticky, tar-like substance when they are heated at low temperatures. This substance makes small coal particles gather into larger lumps which impede handling of the coal and cause it to react unevenly in gasifiers.

ALGAE Fast-growing unicellular or polycellular plants that live in fresh or salt water. They are distinguished from fungi by the presence of chlorophyll and the ability to perform *photosynthesis.*

ALTERNATING CURRENT (A-C) A type of electric current that is produced by impressing a periodic voltage on an electrical system so that the electrical potential is alternately positive and negative. The electrons that conduct the current therefore move alternately in opposite directions with no net displacement. While there is thus no electron flow, the oscillating electrons produce electric fields and heat energy in the same fashion as the unidirectional electron flow of *direct current.* Alternating current is particularly useful for *electric power transmission* because, unlike direct current, its voltage can be raised or lowered readily with static transformers; use of alternating current also simplifies the design and reduces the cost of motors and generators.

ANAEROBIC DIGESTION The use of microorganisms that can live only in the absence of oxygen to decompose organic materials into components that are more valuable or more easily disposed of. The chief use of anaerobic digestion now is reduction of the volume of sewage *sludge,* but it could also be used for the production of *synthetic natural gas* from organic wastes or from *algae.*

BASEBALL A type of *plasma* confinement machine for *fusion* research in which the plasma is held in a linear magnetic bottle sealed at both ends by "magnetic mirrors"—an arrangement of magnetic fields that deflects escaping ions back into the container. The plasma is stabilized by current-carrying structures whose shape resembles the seams of a baseball. The Baseball machines were developed at the University of California's Lawrence Radiation Laboratory at Livermore, California.

BIOCONVERSION A general term describing the conversion of one form of energy into another by plants or microorganisms. Synthesis of organic compounds from carbon dioxide by plants is bioconversion of solar energy into stored chemical energy. Similarly, digestion of solid wastes or sewage *sludge* by microorganisms to form methane is bioconversion of one form of stored chemical energy into another, more useful form.

BREEDER A nuclear *reactor* that produces more fuel than it consumes. Breeding is possible because of two facts of nuclear physics:

1) *Fission* of some atomic nuclei produces more than one neutron for each nucleus undergoing reaction. In simplified terms, then, one neutron can be used to sustain the fission chain reaction and the excess neutrons can be used to create more fuel.

2) Some nonfissionable nuclei can be converted into fissionable nuclei by capture of a *neutron* of proper energy. Nonfissionable uranium-238, for example, can thus be bred into fissionable plutonium-239 in a nuclear reactor powered by plutonium.

A measure of the efficiency of a breeder reactor is the breeding ratio, defined as the number of new fissionable atoms produced per atom of the fissionable species consumed. The practical measure of efficiency, however, is the doubling time, the length of time required for a net doubling of the amount of fissionable material in the reactor core. Most breeders now under consideration have projected doubling times of 10 to 15 years.

Breeder reactors are divided into two types, *fast breeders,* which use high energy neutrons, and *thermal breeders,* which use neutrons of much lower energy.

BRITISH THERMAL UNIT (BTU) The amount of energy necessary to raise the temperature of one pound of water by one degree Fahrenheit at or near 39.2°F (4°C). One Btu is equal to 1054 joules.

CAPACITANCE The property of an electric non-conductor (*dielectric*) that permits storage of energy when opposite surfaces of the conductor are maintained at different electrical potentials. Because the conductors of *electric power transmission* lines are at different potentials, energy is stored in the insulating material that separates them.

CHAR The solid, carbonaceous residue that results from incomplete combustion or *pyrolysis* of organic material. It can be burned for its energy content or, if free from large amounts of impurities, processed further for production of activated carbon for use as a filtering medium. Char produced from coal is generally called coke, while that produced from wood or bone is called charcoal.

COAL A solid, combustible organic material formed by the decomposition of vegetable material without free access to air. Plant debris early in the earth's history accumulated underwater in swamps and gradually decomposed. With the assistance of anaerobic microorganisms, the debris was gradually transformed into peat—partially carbonized vegetable matter. The conversion of peat to coal occurred after most of the water was removed and under conditions of increased pressure and temperature. The conversion, extending over many millions of years, was progressive, leading first to lignite, then to subbituminous, bituminous, and finally to anthracite. Chemically, coal is composed chiefly of condensed aromatic ring structures of high molecular weight. It thus has a higher ratio of carbon to hydrogen content than does *petroleum.*

COAL GAS An artificial gaseous fuel produced by *pyrolysis* of *coal.* It consists primarily of hydrogen, methane, and carbon monoxide, along with some ethane, ethylene, benzene, carbon dioxide, oxygen, and nitrogen, and smaller amounts of aromatic hydrocarbons, water vapor, ammonia, hydrogen sulfide, and other gases. After removal of the noxious components and recovery of the valuable aromatics, coal gas was once fed into local pipelines for use as a domestic fuel. The energy content of this fuel, also known as town gas, is about

450 Btu per standard cubic foot. Use of coal gas in the United States stopped when inexpensive natural gas became widely available.

COMBINED CYCLE POWER PLANT A power plant in which two or more different types of *turbines* are used to extract the maximum amount of useful work from combustion of a fuel. The primary units in such a facility are a gas turbine that extracts energy from the combustion gases before they are used to produce steam and a conventional steam turbine. Additional useful work can also be obtained with a turbine operated by a very high-boiling fluid that extracts energy from the combustion gases before they enter the gas turbine (a topping cycle), or with a turbine operated by a low-boiling fluid that extracts additional energy from the spent steam (a bottoming cycle). The overall efficiency of a combined cycle system may reach 50 percent, compared to the 39 percent efficiency of the best steam turbines now available.

CRACKING The process of producing low molecular weight hydrocarbons from heavier hydrocarbons by heating with or without catalysts. If the reaction is performed in the presence of excess hydrogen, it is called hydrocracking. Both processes are widely used for the production of volatile fuels such as *gasoline.*

CRITICAL MASS The minimum amount of a fissionable material, such as uranium-235 or plutonium-239, that is required to sustain *fission* in a nuclear *reactor.* For ^{239}Pu, the critical mass is about 5 kilograms.

CRUDE OIL *Petroleum* liquids as they come from the ground. Also called simply "crude."

CRYORESISTIVE TRANSMISSION LINES *Electric power transmission* lines in which the conducting cable is cooled to the temperature of liquid nitrogen, 77°K (−196°C). At that temperature, the resistance of the conductor is generally reduced by a factor of about 10, so that less electrical energy is lost in transmission. This saving is offset, however, by the amount of energy needed to cool the cable. The net effect, then, is not greater efficiency of transmission, but increased transmission capacity.

CURIE The unit of *radioactivity.* One curie is the amount of a radioactive *isotope* necessary to produce 3.7×10^{10} disintegrations per second. Uranium-238, for example, disintegrates by emitting an alpha particle (helium nucleus) to become an isotope of thorium. One curie of uranium is thus the amount that is necessary to emit 3.7×10^{10} alpha particles per second.

DELAYED NEUTRONS *Neutrons* produced during the decay of a *fission* product to a lower energy level. Prompt neutrons are those produced during fission itself.

DEUTERIUM An *isotope* of hydrogen in which the nucleus contains a proton and a *neutron.* A deuterium atom is thus about twice as heavy as a hydrogen atom, whose nucleus contains only a proton, but their chemical properties are almost identical. The natural abundance of deuterium, the amount of hydrogen that occurs as deuterium in nature, is about 0.0156 percent. Deuterium is generally obtained by *electrolysis* of deuterium oxide (*heavy water*) that has been separated from normal water by fractional distillation or electrolysis. It is expected to be the primary fuel for *fusion* power plants.

DIELECTRIC A medium through which attraction or repulsion of electrical charges may be sustained. Dielectrics are always insulators. Glass, for example, is a dielectric: if a sheet of glass is placed between two unlike charges, the charges will continue to attract each other. If a conductor such as copper is placed between the charges, however, each will be shielded from the electric field of the other.

DIESEL OIL A *petroleum* fraction composed primarily of aliphatic (linear or unbranched) hydrocarbons. Diesel oil is slightly heavier than *kerosine*, and distills in the range 250 to 400°C. It is used as a fuel in internal combustion engines in which ignition results from high temperatures produced by compression of air, and as a light *fuel oil* (No. 2 fuel oil).

DIRECT CURRENT (D-C) An electric current, such as that produced by a battery, in which the electrical potential does not change its sign, so that the voltage is essentially invariant with time. In a direct current, therefore, energy is carried by a continuous, unidirectional flow of electrons through a conductor. The voltage of direct current cannot be changed with a transformer, thus making it unsuitable for local distribution of electricity. It can, however, be used under some circumstances for *electric power transmission*.

DISTILLATE FUELS The term distillate fuels could correctly be applied to all liquid petroleum fractions other than *residual fuel oil* since all are, in fact, distilled. The most common usage, however, is to describe *diesel oil* and the light *fuel oils* used for residential heating. Distillate fuels, which account for 25 to 30 percent of refined petroleum products, are the fuels that were in shortest supply during last winter's energy squeeze.

ELECTRIC POWER TRANSMISSION Energy may be transmitted through conductors as a flow of electrons under pressure. The flow is measured in amperes, and the pressure or electrical potential is measured in volts. The amount of energy transmitted is proportional to the product of the flow and the pressure. The principal loss of energy in transmission of electricity arises from resistance of the conductor; resistance, in turn, is proportional to the current flow and inversely proportional to the diameter of the conductor. In principle, then, the transmission loss for any given quantity of energy may be reduced by raising the voltage, since that lowers the current flow. Electric power should therefore be transmitted at the highest voltage possible.

At similar voltages, more power can be transmitted by a *direct current* than by an *alternating current*. The power transmitted by a direct current is directly proportional to the voltage; the power transmitted by an alternating current is proportional to the so-called root-mean-square average of the fluctuating voltage, and is thus only about 71 percent as large. Energy transmission by an alternating current, furthermore, is also subject to losses arising from interaction of the current with the insulation surrounding the conducting cable (*capacitance*), and from other factors. Nonetheless, alternating current is the principal method used for transmission of electricity because of the ease with which its voltage may be varied with transformers and because substantial energy losses occur in the *rectification* of alternating current to direct current and in *inversion* of direct current to alternating current.

ELECTROLYSIS The decomposition of a compound into its elements by passage of a *direct current* through an *electrolyte* containing the compound. If the electrolyte is a dilute aqueous solution of alkali salts or bases, then each unit of electricity liberates one volume of oxygen at the positive electrode and two volumes of hydrogen at the negative electrode.

ELECTROLYTE A substance that conducts electricity by the transfer of ions. The most important electrolytes are aqueous solutions of salts, acids, or bases, but some molten salts and a very few solids can also function as electrolytes. Metals are not electrolytes because they conduct electricity by the transfer of electrons.

ELECTRON VOLT (EV) A unit of energy equal to the energy gained by an electron in passing from a point of low potential to a point one volt higher in potential. One million electron volts (1 Mev) is the energy gained by an electron in passing through a potential difference of 1 million volts. One *British thermal unit* is equal to 6.59×10^{21} ev or 6.59×10^{15} Mev.

ELECTROSTATIC PRECIPITATION The use of an electric field to remove solid particles or droplets of liquid from a gas. Electrostatic precipitation is increasingly used by coal-burning power plants to remove *fly ash* from the combustion gases. Precipitation results from interaction of an electric field maintained within the exhaust system and an electric charge induced on the surface of the particle or droplet.

ENRICHMENT The process of increasing the concentration of fissionable uranium-235 in uranium from the naturally occurring level of about 0.7 percent to the concentration required to sustain *fission* in a nuclear *reactor*, generally more than 3 percent. The principal method of enrichment is *gaseous diffusion*, but gaseous centrifugation is also receiving much attention, particularly abroad.

FAST BREEDER A *breeder reactor* that operates with *neutrons* in the "fast" energy range; i.e., with energies greater than 0.1 million *electron volts*. The principal reaction envisioned for most proposed fast breeders is conversion of nonfissionable uranium-238 to fissionable plutonium-239.

FISCHER-TROPSCH GASOLINE A *naphtha*-like artificial *gasoline* produced by combining carbon monoxide and hydrogen over a cobalt-thorium oxide catalyst at 200 to 250°C. The source of the carbon monoxide and some of the hydrogen is generally coal. The Fischer-Tropsch process was developed in Germany before World War II, but has never been used to any great extent because the product is substantially more expensive than gasoline produced from petroleum.

FISSION The splitting of certain heavy atomic nuclei into two atoms of much lower atomic weight and whose aggregate mass is less than that of the parent nucleus. The process is initiated by the capture of a *neutron* by the nucleus of the fissionable atom, and is accompanied by the emission of one to three new neutrons. The lost mass appears as energy (E) in the amount $E = mc^2$, where m is the change in mass and c is the speed of light (about 186,000 miles per second). The energy released in fission is much greater than that released in simple radioactive decay and, since fission produces neutrons, the reaction can be made self-perpetuating. Self-perpetuation, the initiation of fission in adjacent nuclei by neutrons from a nucleus that has undergone fission, is known as a chain reaction. If the chain reaction proceeds slowly, as when some neutrons are prevented from hitting adjacent fissionable nuclei by the presence of a *moderator,* it produces heat that can be used for production of steam to generate electricity. If the chain reaction proceeds too rapidly, it produces an explosion of tremendous force.

FLUIDIZED BED A reaction chamber in which finely divided solid reactants are suspended and maintained in a state of turbulent motion by a stream of gas or liquid from below. As a result, the reactants flow and mix freely, the entire surface area of the particles is exposed to the fluid for reaction, and a high rate of heat transfer is obtained.

FLY ASH The fine, solid particles of noncombustible mineral residue carried from a bed of solid fuel by the gaseous products of combustion.

FUEL CELL A battery or electrochemical cell in which the chemicals that combine to produce *direct current* electricity do not remain in the cell after the reaction. The simplest fuel cell is composed of a fuel electrode (anode), an oxidant or air electrode (cathode), and an *electrolyte*. Liquid or gaseous fuel is fed to the anode where it is catalytically oxidized, releasing electrons to an external circuit. At the cathode, these electrons reduce the oxidant to ions, which migrate through the electrolyte and combine with the reduced fuel to form the oxidation product, which is then removed from the cell. The net result is direct production of electricity from combustion of the fuel. Fuel cells produce electricity with nearly twice the efficiency of conventional generators.

FUEL OILS Fuel oils are the *petroleum* fractions with a higher boiling range than *kerosine*. They are generally classified as either distillates or residuals. Distillates (Nos. 1, 2, and 4) are the lighter oils used primarily for central heating of homes, small apartment houses, and commercial buildings, and for transportation. Residuals (Nos. 5 and 6), often called bunker oils, are heavier, high-viscosity oils that usually need to be heated before they can be pumped and handled conveniently. They are used primarily in industry, in large commercial buildings, and for the generation of electricity.

FUEL PELLET For *laser-induced fusion* power plants, the fuel is a small pellet of frozen *deuterium* and *tritium*. The optimum size for such a pellet has not yet been determined; current proposals range from 1 millimeter in diameter to 1 centimeter.

FUEL REPROCESSING The chemical or metallurgical treatment of used fuel from a nuclear *reactor* for the purpose of recovering and decontaminating fissionable materials. The principal operations involved in reprocessing are:

1) Decay cooling, in which the spent fuel is stored for three to six months, often under water, to allow for decay of short-lived *fission* products.
2) Removal of the fuel element cladding and dissolution of the fuel and its support material.
3) Chemical separation of the fissionable and fertile constituents. (Fertile atoms are those that can be converted to fissionable atoms in a *breeder reactor.*)
4) Product recovery.
5) Disposal of radioactive wastes.

Because of the intense *radioactivity* of the spent fuel, most of the operations of a reprocessing plant must be performed behind massive shielding and by remote control. Fuel reprocessing plants are thus much more expensive than conventional chemical plants of comparable size and complexity.

FUSION The combination of two atomic nuclei to yield one larger nucleus whose mass is less than the aggregate mass of the original nuclei; the lost mass appears as energy in the same manner as in *fission*. Energy production in both fission and fusion is related to the stability of the atomic nucleus, and iron has the most stable nucleus of all the elements. If two small atomic nuclei are combined to produce a nucleus whose mass is closer to that of iron, the nucleus that is formed will be more stable and energy will be released by fusion. Similarly, if an atom much larger than iron is broken down into two smaller atoms whose nuclei are only slightly heavier than iron, the new nuclei are also more stable and energy is released by fission. A net energy input is thus required for formation of atoms heavier than iron or for fission of atoms lighter than iron.

The most important reactions considered for a fusion power plant are:

1) The combination of two *deuterium* nuclei (one proton and one neutron apiece) to produce a helium-3 nucleus (two protons and one neutron), a neutron, and 3.2 million *electron volts* of energy.

2) The combination of two deuterium nuclei to produce one *tritium* nucleus (one proton and two neutrons), one proton, and 4.0 Mev of energy.

3) The combination of a deuterium nucleus and a tritium nucleus to produce a helium-4 nucleus (two protons and two neutrons), one neutron, and 17.6 Mev of energy. Because of its high energy release and because it can be initiated at a lower temperature, this reaction is most often proposed as the basis of fusion power plants.

Electrical charges and atomic forces make it very difficult to bring the nuclei close enough together for fusion to occur. Initiation of the reaction therefore requires a combination of very high temperatures and pressures, much higher than have ever been produced under controlled conditions. Because of this requirement for heat, fusion is often referred to as a thermonuclear reaction. The hydrogen or thermonuclear bomb is an uncontrolled, explosive fusion reaction in which the necessary high temperature is achieved by detonation of an atomic bomb.

GASEOUS DIFFUSION The principal process for *enrichment* of uranium; that is, for increasing the concentration of fissionable uranium-235 in a mixture of uranium *isotopes* to the level required to sustain *fission* in a nuclear *reactor*. Uranium is converted to gaseous uranium hexafluoride, which is then forced through a long series of porous barriers. Since ^{235}U is slightly lighter than the more abundant ^{238}U, the uranium hexafluoride containing ^{235}U will pass through the barriers at a slightly faster rate, and the first gas to pass all the way through the barriers will be enriched in it.

Uranium enrichment requires very large, very expensive facilities and great quantities of electricity. Nuclear power plants now in operation in the United States produce only slightly more electricity than is consumed in gaseous diffusion plants, but this comparison is misleading because much of the enriched uranium produced in these plants is sold abroad.

GASIFICATION In the most commonly used sense, gasification refers to the conversion of coal to a high-Btu *synthetic natural gas* under conditions of high temperatures and pressures. It can, however, also refer to conversion of coal to a low-Btu *power gas* for on-site generation of electricity, or the conversion of any organic material to either of these products.

GASOLINE A *petroleum* fraction composed primarily of small branched chain, cyclic, and aromatic hydrocarbons boiling in the range 60 to 200°C. Since these light hydrocarbons form only a small percentage of crude petroleum, a substantial portion of gasoline is produced by *cracking, hydroforming,* and *reforming* heavier hydrocarbons during the *refining* process.

HEAT PUMP A device which transfers heat from a colder to a hotter reservoir by the expenditure of mechanical (or electrical) energy when the primary purpose is heating the hot reservoir rather than refrigerating the cooler one. A heat pump is essentially a reversed refrigeration process. In a refrigeration system, the refrigerant is compressed, the resulting heat is removed by radiation from a condenser, and the refrigerant is then allowed to expand to cool (absorb heat from) the stored material. The heat pump is based on the same system, but a different part of the cycle is used for the useful function. Air or water drawn from an outside source is cooled by having the refrigerant absorb its heat. The refrigerant is then compressed and the heat is removed at the condenser and used for heating. Heat pumps are a far more efficient method of electric residential heating than the resistive heating now commonly used.

HEAVY WATER Deuterium oxide; that is, water in which all hydrogen atoms have been replaced by *deuterium*. The molecular weight of heavy water is thus 20, compared to 18 for ordinary water. Heavy water's melting point is 3.8°C, its boiling point is 101.42°C, and its density at 25°C is 1.1056 times that of ordinary water. Deuterium oxide is generally separated from ordinary water by fractional distillation; it is also concentrated in *electrolysis* cells that have operated for a long time, since it decomposes less readily than ordinary water. Its principal use is as a *moderator* in nuclear power plants.

HYDROFORMING The process of converting low-octane hydrocarbons into a stable fuel of higher octane rating by heating them under pressure in the presence of hydrogen and catalysts. The principal reactions are conversion of cyclic hydrocarbons into aromatics by loss of hydrogen, conversion of cyclic hydrocarbons into linear hydrocarbons of lower boiling point by *cracking,* and conversion of complex linear hydrocarbons into aromatic and linear hydrocarbons of lower boiling point. Hydroforming is an important operation in the production of *gasoline.*

HYDROFRACTURING The use of water and water-based solutions to form underground channels and free trapped petroleum. Fluid pumped into a producing oil well under great pressure splits open horizontal fissures in the reservoir rock. Propping agents such as sand or pellets are added to the water to hold the fissures open when the pressure is released.

HYPERSONIC Referring to speeds greater than five times the speed of sound in air. The term supersonic refers to speeds of one to five times the speed of sound in air.

INTEGRAL TRAIN A type of train proposed, but never used, for transportation of coal or other minerals from the mine to a single point of consumption. It is similar to a *unit train,* except that it would be composed of special cars permanently linked together with couplers that would allow each car to be emptied by a rotary dump that would invert the whole car.

INVERSION The conversion of *direct current* electricity to *alternating current.* A typical inverter uses electron tubes or transistors to cause direct current to flow in alternating directions through the primary coil of a transformer. This flow induces an alternating current in the secondary coil of the transformer. By varying the size of the transformer, alternating current can be obtained in a range of voltages.

ISOTOPE Any of two or more species of atoms with the same atomic number (the same number of protons in the nucleus), but with different atomic masses because of differing numbers of *neutrons* in the nucleus. All isotopes of an element have the same number of orbital electrons, and thus very similar chemical properties, but the differing atomic masses produce slightly different physical properties. Since the atomic mass governs the stability of the nucleus, one or more isotopes of an element may be radioactive or fissionable while other isotopes of the same element are stable.

KEROGEN A resinous hydrocarbon material, of general formula $(C_6H_8O)_n$, that is the chief organic constituent of *oil shale.* When heated to 450 to 600°C, kerogen releases vapors that can be converted to raw shale oil, a black, viscous mixture of hydrocarbons. Shale oil can, in turn, be converted into petroleum products by *refining.*

KEROSINE The petroleum fraction containing hydrocarbons that are slightly heavier than those found in *gasoline* and *naphtha,* with a boiling range of about 180 to 300°C. Prior to 1910, kerosine (also spelled kerosene) was the most important petroleum product because of its use for home and commercial lighting. Demand fell sharply after the introduction of electric lights, but has risen in recent years as a result of kerosine's use in gas turbines and jet engines.

KILOWATT (KW) The unit of power equal to 1000 *watts.*

KILOWATT-HOUR (KWHR) A unit of work or energy equivalent to the steady consumption of one *kilowatt* of power for a period of one hour.

LASER-INDUCED FUSION A proposed concept in which the high temperature and pressure required for initiating fusion in a *plasma* are produced by bombarding *fuel pellets* with short, intense bursts of radiation from one or more lasers. Theoretical calculations suggest that inertial forces could contain the plasma produced

by each burst long enough for fusion to occur; the result is a series of small thermonuclear explosions. Energy released in these explosions could be absorbed in a heat sink of molten lithium or lithium salts and used for production of steam and generation of electricity. Laser-induced fusion is still only a theoretical concept, however, since there are as yet no lasers that can deliver enough energy to the pellets in a sufficiently short time.

LAWSON CRITERION The basic theoretical requirement for operation of a *fusion* power plant, determined by British physicist J. D. Lawson. For a *deuterium-tritium* fuel mixture in the temperature range 100 million to 500 million degrees Centigrade, Lawson found that the product of the ionic density in a fusion *plasma* and the confinement time must be about 10^{14} seconds per cubic centimeter for the amount of energy produced by fusion to equal that required to maintain the confined plasma (assuming a 33 percent efficiency in converting heat to electricity). This figure thus represents the break-even point beyond which the fusion reaction gives net production of energy.

LIQUEFIED PETROLEUM GAS (LPG) Also known as "bottled gas," liquefied petroleum gas consists primarily of propanes and butanes recovered from *natural gas* and in the *refining* of petroleum. It has an energy content of 2000 to 3500 Btu per standard cubic foot. LPG is widely used as a fuel for internal combustion engines in applications where pollution must be minimized, such as in buildings and mines, but the largest use is as a substitute for natural gas in areas not served by pipelines.

LIQUEFIED NATURAL GAS (LNG) *Natural gas* that has been cooled to about −160°C for shipment or storage as a liquid. Liquefaction greatly reduces the volume of the gas, and thus reduces the cost of shipment and storage, even though high pressure cryogenic containers must be used.

MAGNETOHYDRODYNAMIC (MHD) GENERATOR An expansion engine in which hot, partially ionized gases are forced through a magnetic field. Movement of the electrically conducting gas through the field generates an electric current that is collected by electrodes lining the expansion chamber. MHD is thus an efficient way to generate electricity from the combustion of fuels without going through an intermediary steam *turbine.* If exhaust gases from the expansion chamber are used to operate a steam turbine, the efficiency of fuel use is further increased. MHD is potentially a valuable method for burning *coal* with a high sulfur content, since the sulfur combines with ionized gases in the expansion chamber and thus can be precipitated readily.

MEGAWATT (MW) A unit of power equal to 1000 kilowatts or 1 million *watts.*

MODERATOR A substance used to slow *neutrons* to a speed at which there is a higher probability of initiating *fission* in a nuclear *reactor.* The neutrons lose energy by colliding with molecules of the moderator. The most commonly used moderators include graphite, water, *heavy water,* and beryllium.

NAPHTHA A loosely defined petroleum fraction containing primarily aliphatic (linear) hydrocarbons with boiling points ranging from 125 to 240°C. It is thus intermediate between *gasoline* and *kerosine,* and contains components of both. Its principal uses are in solvents and paint thinners and as a raw material for the production of organic chemicals, but it is expected to be used increasingly as a raw material for the production of *synthetic natural gas.*

NATURAL GAS A gaseous fossil fuel generally found in association with oil and whose composition varies with its origin. Most unprocessed natural gases contain about 60 to 80 percent methane, 5 to 9 percent ethane, 3 to 18 percent propane, and 2 to 14 percent heavier hydrocarbons. Varying quantities of nonhydrocarbon gases are usually also present, including nitrogen, carbon dioxide, and hydrogen sulfide. The energy content of raw natural gas varies from 900 to 1300 Btu per standard cubic foot.

Propane and heavier hydrocarbons are usually removed by liquefication and sold as *liquefied petroleum gas* (natural gas liquids) or separated and sold individually. After removal of hydrogen sulfide, pipeline natural gas generally contains 70 to 90 percent methane, 6 to 24 percent ethane, and 1 to 8 percent propane, and has an energy content of about 980 to 1050 Btu per standard cubic foot.

NEUTRON An uncharged (neutral) elementary particle with a mass slightly larger than that of a proton. It is present in all atomic nuclei except that of hydrogen. Because it has no electrical charge, the neutron is able to penetrate the dense negatively charged electron cloud of an atom and interact with the strongly positive nucleus. Neutrons are not emitted in radioactive decay of atoms, but are released in nuclear reactions such as *fission* and *fusion.*

NEUTRON FLUX DENSITY Also called neutron flux. The number of *neutrons* that enter a sphere of unit cross-sectional area per unit time. Less rigorously, it may also be defined as the number of unidirectional neutrons that pass in unit time through a unit area perpendicular to the neutron beam. Neutron flux density is one of the most important factors that must be considered in design of a nuclear reactor because a minimum flux is necessary to sustain *fission* and high fluxes damage reactor materials.

NUPLEX A large industrial complex centered around and dependent upon a nuclear power plant. In nuplexes, waste heat produced in the generation of electricity would be used as process heat by the industries, thereby greatly increasing the efficiency of nuclear fuel use.

OIL SHALE A fine-grained, laminated sedimentary rock that contains an oil-yielding organic material called *kerogen.* In the U.S., oil shale is found primarily in Colorado; much larger deposits exist in several other countries. Upon heating, oil shales yield from 12 to 60 gallons of oil per ton of rock, although the best U.S. oil shale produces less than 30 gallons per ton.

PETROLEUM Crude oil and *natural gas,* although the term is frequently used to refer only to the oil. Petroleum was formed from animal and vegetable material that collected at the bottom of ancient seas. This debris was buried by inorganic residue and subjected to great heat and pressure, and perhaps to bacterial and chemical action. Petroleum contains a much greater concentration of lower molecular weight hydrocarbons and a higher ratio of hydrogen to carbon than does *coal.*

PHOTOLYSIS The decomposition of molecules caused by the absorption of light energy. When a molecule of water, for example, is exposed to light of the proper wavelength, it breaks apart into hydrogen and oxygen.

PHOTON The smallest unit (quantum) into which electromagnetic radiation can be divided. The amount of energy contained in a photon is directly proportional to the frequency of the electromagnetic radiation.

PHOTOSYNTHESIS The reaction by which green plants convert carbon dioxide into more complex organic materials, particularly carbohydrates such as glucose. Energy for the reaction is provided by absorbed sunlight, and the reaction is catalyzed by chlorophyll. Photosynthesis is the primary method for *bioconversion* of solar energy into forms more useful to man.

PHOTOVOLTAIC CELL A type of semiconductor in which the absorption of light energy creates a separation of electrical charges. This separation creates an electrical potential that can be tapped by allowing electrons to flow through an external circuit. The net effect is direct conversion of light, especially solar energy, into electricity. The efficiency of such cells is generally very low, however, and their cost is still quite high. Typical materials used in the construction of photovoltaic cells are silicon, cadmium sulfide, and gallium arsenide.

PILOT PLANT A small plant used to test or develop a new manufacturing process. The chief function of a pilot plant is to test the technological and economic feasibility of scaling a laboratory process up to industrial size. The National Science Foundation prefers to refer to pilot plants as proof-of-concept plants. The capacity of a pilot plant is generally about 1 to 10 percent that of a full scale unit. Demonstration plants are generally intermediate in size between pilot and full scale plants, and are usually used when a larger scale operation is necessary to assess all aspects of a new process.

PLASMA An electrically neutral, partially ionized gas in which the motion of the constituent particles is dominated by electromagnetic interactions. The study of these interactions is called magnetohydrodynamics (MHD).

PNEUMOCONEOSIS A crippling lung disease, marked by large amounts of fibrous growth, that results from the inhalation of dust. When the dust is from coal, the condition is often called black lung disease.

POWER The rate at which work is done or energy is transferred. Power is measured in units of work (such as the *British thermal unit*) per unit time; typical units are the horsepower and the *watt*.

POWER GAS A low-energy gaseous fuel, generally produced from *coal,* whose principal combustible components (carbon monoxide and hydrogen) are greatly diluted by air. Also called producer gas, it differs from *coal gas* in that the energy content is only about 150 to 250 Btu per standard cubic foot. Because of its low energy content, power gas cannot be transported economically and is generally burned at the production site for the generation of electricity.

PUMPED HYDROELECTRIC STORAGE The only method now in commercial use for large scale storage of electricity. Excess electricity produced at times of low demand is used to pump water into a reservoir. At times of high demand, this water is released and used to operate hydroelectric generators to produce supplemental electricity. Pumped hydroelectric storage returns only about two thirds of the electricity put into it, but it costs less than an equivalent amount of additional generating capacity.

PYROLYSIS The transformation of a substance into another compound or compounds by the application of heat alone. In the context of energy, pyrolysis (also called destructive distillation) is the heating of organic materials such as *coal,* wood, *petroleum,* and solid wastes in the absence of oxygen with provision for recovery of the desired combustible products. If heat is applied slowly, the initial products are water vapor and volatile organic compounds. Increased heat leads to recombination of the organic materials into complex hydrocarbons and water. The principal products of pyrolysis are gases, oils, and a solid residue called *char*. It is possible to produce great variations in the relative proportions of these products by varying the pyrolysis conditions.

RADIOACTIVITY The spontaneous disintegration of the nucleus of an atom with the emission of corpuscular or electromagnetic radiation. These emissions are of three principal types, called alpha, beta, and gamma. Alpha radiation is composed of positively charged helium nuclei (two protons and two *neutrons*) ejected with a velocity 5 to 7 percent that of light. Beta radiation is composed of negative electrons ejected with velocities which may approach the speed of light. Gamma radiation is uncharged electromagnetic radiation similar to x-rays. Gamma radiation is by far the most dangerous of the three, since its penetrating

power is approximately 100 times that of beta radiation and about 10,000 times that of alpha radiation.

REACTOR An assembly of nuclear fuel capable of sustaining a *fission* chain reaction. A reactor is sometimes also called an atomic pile, since the first reactor—built during World War II for the Manhattan Project—was composed of a pile of alternating layers of uranium (or uranium oxide) and a graphite *moderator*.

RECTIFICATION The conversion of *alternating current* to *direct current*. In general, rectifiers are devices that allow current to flow in only one direction. In the thermionic rectifier, for example, a hot cathode emits electrons that are attracted to another electrode only when it is positively charged; application of an alternating current to this electrode produces alternating positive and negative charges. Since current can pass only when the electrode is positive, the effect is conversion of alternating current to direct current. Most modern rectifiers are semiconductor devices that accomplish the same thing in a more elegant fashion.

REFINING The separation of *petroleum* into distinct fractions to obtain useful products. Refining consists primarily of distilling the petroleum into fractions of different boiling ranges, converting the less valuable components into more valuable products (*reforming*), and treating the fractions to remove undesirable components. The main fractions are gas, *gasoline, naphtha, kerosine,* gas oils, and *residual fuel oil*.

REFORMING The process of converting low-octane hydrocarbons into a stable fuel of higher octane rating by heating them under pressure in the presence of catalysts. If hydrogen is also present, the process is called *hydroforming*.

RESERVES Known deposits of minerals or fossil fuels that can be removed from the earth at a reasonable cost.

RESIDUAL FUEL OIL A high-viscosity *fuel oil* that must be heated before it can be pumped and handled conveniently. Residual fuel oil is the petroleum fraction that is collected after all lower-boiling fractions have been distilled away. It is used primarily in industry, in large commercial buildings, and for the generation of electricity.

SCYLLAC A *plasma* confinement machine for *fusion* research in which the magnetic field is in the shape of a *toroid*. Scyllac is a representative of the so-called "theta-pinch" machines in which a secondary plasma-stabilizing magnetic field in the axial direction is generated by an electric current flowing around the axis of the toroid. Scyllac machines were developed at the Los Alamos Scientific Laboratory, Los Alamos, New Mexico.

SLUDGE The solid organic material that separates from sewage held in settling tanks. Partial oxidation of protein in the sludge makes it putrid and foul-smelling within a short time.

It is therefore generally further treated by *anaerobic digestion,* which produces considerable liquefaction and reduction in volume. The digested sludge is relatively inoffensive, and may be disposed of by being dried and spread on the earth, although it has little value as fertilizer. It does have a high protein content, however, and some investigators are considering its use as a food for animals.

SOLVENT REFINING Also called hydroprocessing or hydrofining. A method for removing ash and sulfur compounds from *coal.* The combustible organic portion of the coal is dissolved in an organic solvent, generally one that has been made from the coal itself. The solution is filtered to remove ash and insoluble organic materials, and fractionated to recover the solvent. The principal product is a heavy organic material called solvent-refined coal; this material melts at temperatures above 180°C, has an energy content of about 16,000 Btu per pound regardless of the source of the coal, and can be used as a substitute for coal in most applications.

STANDARD CUBIC FOOT (SCF) The amount of gas contained in a volume of one cubic foot under standard conditions of temperature and pressure. In physics, the standard pressure is one atmosphere (14.73 pounds per square inch) and the standard temperature is 0°C (32°F). In referring to natural gas, however, the standard temperature is generally taken as 15.6°C (60°F).

SUPERCONDUCTIVITY The disappearance of resistance to an electric current that occurs abruptly when certain metals or alloys are cooled to temperatures near absolute zero (0°K or −273.16°C). Metals that exhibit this behavior include mercury, magnesium, zinc, cadmium, aluminum, lead, and niobium. The highest temperature at which superconductivity has been observed is 21°K (−252.2°C) for an alloy of niobium, aluminum, and germanium.

SYNTHESIS GAS A mixture of approximately equal parts of hydrogen and carbon monoxide formed by reacting steam with hot *coal* or *char.* The mixture can be burned for fuel, and is thus similar to *coal gas,* but the primary use is in the production of methane or *synthetic natural gas.* Synthesis gas, also known as water gas, has an energy content of about 250 to 350 Btu per standard cubic foot.

SYNTHETIC NATURAL GAS (SNG) A manufactured gaseous fuel generally produced from *naphtha* or *coal.* It contains 95 to 98 percent methane, and has an energy content of 980 to 1035 Btu per standard cubic foot, about the same as that of *natural gas.*

THERMAL BREEDER A *breeder reactor* that operates with *neutrons* in the thermal energy range; that is, neutrons with energies less than 1 *electron volt.* The reaction most often considered for use in thermal breeders is conversion of nonfissionable thorium-232 into fissionable uranium-233.

TOKAMAK A *plasma* confinement machine for *fusion* research in which the magnetic field is in the shape of a *toroid.* In the tokamak, a secondary plasma-stabilizing magnetic field is generated by an electric current flowing axially through the plasma itself. The tokamak is thought by many to be the most promising approach to magnetically confined fusion. The largest U.S. tokamak is located at the Plasma Physics Laboratory of Princeton University, Princeton, New Jersey.

TOROID A doughnut-shaped surface generated geometrically by rotation of a circle around an axis that lies in the plane of the circle but does not intersect the circle. The toroid is the shape of the magnetic container that will confine the *plasma* in some proposed *fusion* power plants.

TRITIUM An *isotope* of hydrogen in which the nucleus contains a proton and two *neutrons.* A tritium atom is thus about three times as heavy as a hydrogen atom, whose nucleus contains only a proton. Tritium is radioactive, disintegrating to a helium-3 nucleus by emission of an electron. Its half-life is about 12.4 years; that is, half of any quantity of tritium will decay to helium-3 in 12.4 years. The natural abundance of tritium is about 1 part tritium in 10^{17} parts of hydrogen. It is produced in nature by cosmic rays in the upper atmosphere, but it can be produced artificially in a nuclear *reactor.* The principal use of tritium is as a radioactive tracer for studying chemical reactions, but it is expected to be used as a fuel in *fusion* power plants.

TURBINE A rotary engine turned by the impulse from a current of fluid under pressure. A turbine is usually made with a series of curved vanes on a central rotating spindle. Simple examples of a turbine are a windmill and a waterwheel.

ULTRAHIGH VOLTAGES (UHV) Voltages greater than 765,000 volts, the highest voltage now used for *electric power transmission* in the United States.

UNIT TRAIN A string of locomotives and cars used exclusively for bulk shipment of minerals or coal from the mine to a single point of consumption. Because all the cars go to one destination, the locomotives can be distributed more efficiently throughout the train, there is no expense of assembling the train, and the overall cost of operation is substantially lower than for a conventional train.

WATT The amount of work available from an electric current of 1 ampere at a potential of 1 volt. The watt is also the metric unit of *power,* and is equal to a rate of energy consumption of 1 joule per second. One joule is roughly one thousandth of a *British thermal unit;* 1000 watts (1 kilowatt), therefore, is roughly the amount of energy required to raise the temperature of one pound of water one degree Fahrenheit per second.

BIBLIOGRAPHY

The entries in the following bibliography represent key sources of information about subjects discussed in each chapter. The entries are not meant to be comprehensive, but rather to provide a starting point for readers seeking further information. Most of the reports cited list additional references.

Proceedings from the 7th Intersociety Energy Conversion Engineering Conference (IECEC) are no longer available. Reprints of individual papers may, however, be available from the authors.

Cited government documents may be ordered from the appropriate agency by enclosing payment and listing both the title and the catalog number of the report desired. The addresses are:

Superintendent of Documents
U.S. Government Printing Office
North Capitol and H Streets, N.W.
Washington, D.C. 20401

U.S. Department of Commerce
National Technical Information Service
5285 Port Royal Road
Springfield, Virginia 22151
Attention: Storage and Dissemination
Section

GENERAL BIBLIOGRAPHY

1. D. C. White et al., *Energy Technology to the Year 2000,* Technology Review, Cambridge, Massachusetts, 1972.
2. Special Issue on Energy and Power, *Scientific American* **224**, No. 3 (1971).
3. Committee on U.S. Energy Outlook, *United States Energy Outlook,* National Petroleum Council, Washington, 1973.
4. Council on Economic Priorities, "The Price of Power," Economic Priorities Report **3**, No. 2, 1972. 456 Greenwich St., New York, N.Y. 10013.
5. U.S. Geological Survey, "Energy Resources of the United States," Circular 650, Washington, 1972. Available without charge.
6. E. Hirst and D. Steiner, Eds., *Issues Associated with the Use of Energy: Toward a National Energy Policy,* Oak Ridge National Laboratory, Report ORNL-CF-72-8-4, Oak Ridge, Tennessee, 1972.

CHAPTER 1

1. *Mineral Facts and Problems, 1970,* Bureau of Mines Bulletin 650, Washington, 1970, $10.75.
2. *Minerals Yearbook, 1970,* Volume I, Bureau of Mines, Washington, 1971, $6.00.
3. J. A. DeCarlo, E. T. Sheridan, and Z. E.

Murphy, "Sulfur Content of United States Coals," Bureau of Mines Information Circular 8312, Washington, 1966, $0.30.

4. National Academy of Engineering, "Final Report of the Advisory Committee on Project Gasoline," Government Printing Office number I63.10:62, Washington, 1971, $0.30.

5. Foster Wheeler Corporation, "Engineering Evaluation of Project Gasoline, Consol Synthetic Fuel Process," Government Printing Office number I63.10:59, Washington, 1971, $2.75.

6. FMC Corporation, "Char Oil Energy Development—The Desulfurization of COED Char, Part III," Government Printing Office number I63.10:56/INT 2, Washington, 1971, $1.25.

7. FMC Corporation, "Char Oil Energy Development," Government Printing Office number I63.10:56/INT 1, Washington, 1970, $2.50.

8. Pittsburgh & Midway Coal Mining Company, "Economic Evaluation of a Process to Produce Ashless, Low-Sulfur Fuel from Coal," Government Printing Office number I63.10:53, Washington, 1970, $2.50.

9. Pittsburgh & Midway Coal Mining Company, "Development of a Process for Producing an Ashless, Low-Sulfur Fuel from Coal—Design of a Pilot Plant," Government Printing Office number I63.10:53/INT 2, Washington, 1971, $3.50.

10. Ralph M. Parsons Company, "1969 Feasibility Report—Consol Synthetic Fuel Process—Synthetic Crude Production," National Technical Information Service number PB–184330, Springfield, 1969, $6.00.

11. National Economic Research Associates, Inc., "Fuels for the Electric Utility Industry 1971–1985," Edison Electric Institute, New York, 1972.

CHAPTER 2

1. Office of Coal Research, *Annual Report 1972*, Government Printing Office number 2414-0049, Washington, 1972, $1.00. The best reference available.

2. D. Hebden and G. Percival, "New Horizons for Pressure Gasification," *Journal of the Institution of Gas Engineers* 12, No. 8, 1 (1972).

3. J. P. Henry, Jr., B. M. Louks, and S. B. Alpert, "Gasification of Bituminous Coal —Status and Prospects," presented at Joint Power Engineering Conference, Boston, September 13, 1972. Stanford Research Institute, Menlo Park, California 94025.

4. Consolidation Coal Company, "Low-Sulfur Boiler Fuel Using the Consol CO_2 Acceptor Process," National Technical Information Service number PB-176910, Springfield, 1967, $6.00.

5. Bituminous Coal Research, Inc., "Gas Generator Research and Development Phase II—Process and Equipment Development," Government Printing Office number I63.10:20, Washington, 1971, $10.25.

6. Institute of Gas Technology, "Cost Estimate of a 500 Billion Btu/day Pipeline Gas Plant via Hydrogasification and Electrothermal Gasification of Lignite," National Technical Information Service number PB-193928, Springfield, 1968, $6.00.

7. M. W. Kellogg Company, "Commercial Potential for the Kellogg Coal Gasification Process," National Technical Information Service number PB-180358, Springfield, 1968, $6.00.

8. Air Products and Chemicals, Inc., "Engineering Study and Technical Evaluation of BCR 'Two-Stage Super-Pressure Gasification Process'," Government Printing Office number I63.10:60, Washington, 1971, $2.25.

9. J. P. Henry, Jr., and B. M. Loucks, "An Economic Study of Pipeline Gas Production from Coal," *Chemical Technology* 1, 238 (1971).

CHAPTER 3

1. F. L. Robson, A. J. Giramonti, G. P. Lewis, and G. Gruber, "Technological and Economic Feasibility of Advanced Power Cycles and Methods of Producing Nonpolluting Fuels for Utility Power Stations," prepared for the National Air Pollution Control Administration, U.S. Department of Health, Education, and Welfare, Durham, North Carolina 27701 by the United Aircraft Research Laboratories, East Hartford, Connecticut 06108, 1970.

2. A more condensed and readable report that is largely derived from Reference 1 is: A. J. Giramonti, "Advanced power cycles for Connecticut electric utility stations," Report L-971090-2 prepared for the Connecticut Development Commission by United Aircraft Research Laboratories, East Hartford, Connecticut 06108, 1972.

3. L. O. Tomlinson, "Comparison of Combined Cycle Plants Available Today and Future Trends," paper presented at the Joint Power Generation Conference, Boston, Massachusetts, Sept. 10–14, 1972. Power Systems Engineering, Inc., Houston, Texas.

4. S. B. Alpert, R. H. Wolk, and A. M. Squires, *Power Generation and Environmental Change,* M. I. T. Press, Cambridge, Massachusetts, 1971.
5. Arthur M. Squires, "Capturing Sulfur During Combustion," *Energy Technology to the Year 2000,* Technology Review, Cambridge, Massachusetts, 1972.
6. A. N. Dravid, C. J. Kuhre, and J. A. Sykes, Jr., "Power Generation Using the Shell Gasification Process," Proceedings of the Third International Conference on Fluid-ized Bed Combustion, Hueston Woods, Ohio, October 29-November 1, 1972. Proceedings to be published by the Environmental Protection Agency.
7. D. H. Archer, E. J. Vidt, and J. P. Morris, "Coal Gasification for Clean Power Production," Proceedings of the Third International Conference on Fluidized Bed Combustion, Hueston Woods, Ohio, October 29-November 1, 1972, Volume 2, 335. To be published by the Environmental Protection Agency.

CHAPTER 4

1. J. B. Dicks, "MHD Central Power: A Status Report," *Mechanical Engineering* **94,** No. 5, 14 (1972).
2. D. Bienstock, R. J. Denski, and J. J. Demeter, "Environmental Aspects of MHD," Proceedings of the 64th Intersociety Energy Conversion Engineering Conference, New York, 1971. Bureau of Mines Energy Research Center, Pittsburgh, Pennsylvania.
3. MHD Power Generation Study Group, *Open Cycle Coal Burning MHD Power Generation: An Assessment and a Plan for Action,* Government Printing Office, Washington, 1971.
4. R. J. Rosa, *MHD Energy Conversion,* McGraw Hill, New York, 1968.
5. P. Sporn and A. R. Kantrowitz, "Large Scale Generation of Electric Power by Application of the Magnetohydrodynamic Concept," *Power* **103,** 62 (1959).
6. R. J. Rosa, "Physical Principles of Magnetohydrodynamic Power Generation," *Physics of Fluids* **4,** 182 (1961).
7. D. Bienstock et al., "Air Pollution Aspects of MHD," Proceedings of the 13th Symposium on the Engineering Aspects of Magnetohydrodynamics, March 26–28, 1973, Stanford University, Stanford, California. Bureau of Mines Energy Research Center, Pittsburgh, Pennsylvania.
8. V. A. Kirillin et al., "Investigations at U-02 MHD Plant—Some Results," Proceedings of the Fifth International Conference on Magnetohydrodynamical Electrical Power Generation **1,** 353, Organization for Economic Cooperation and Development, Munich, Germany, 1971.
9. K. Fushimi et al., "Development of a Long Duration MHD Channel," Proceedings of the Fifth International Conference on Magnetohydrodynamical Electrical Power Generation **1,** 371, Organization for Economic Cooperation and Development, Munich, Germany, 1971.

CHAPTER 6

1. U.S. Atomic Energy Commission, "Civilian Nuclear Power," Government Printing Office, Washington, 1967. $0.40.
2. A. M. Weinberg, "The Moral Imperatives of Nuclear Energy," *Nuclear News* **14,** No. 12, 33 (1971); *Science* **177,** 27 (1972).
3. Testimony presented to the AEC Public Rule-Making Hearings on Interim Acceptance Criteria for Emergency Core-Cooling Systems, January 27 to November 1, 1972, Bethesda, Maryland (transcripts may be inspected at the AEC Public Document Center in Washington, D.C.).
4. D. H. Ford and H. W. Kendall, *Environment* **14,** No. 7, 2 (1972).
5. H. Alfven, *Bulletin of the Atomic Scientists* **28,** No. 5, 5 (1972).
6. C. E. Larson, "International Economic Implications of the Nuclear Fuel Cycle," lecture at the Symposium on Energy, Resources and the Environment, Kyoto, Japan, July 10, 1972. U.S. Atomic Energy Commission, Washington, 1972.
7. J. P. Rossie and W. A. Williams, Jr., "The Cost of Energy from Nuclear Power Plants Equipped with Dry Cooling Systems," paper presented at the Joint Power Generation Conference, Boston, Massachusetts, September 10–14, 1972. American Society of Mechanical Engineers, New York.
8. W. E. Johnson, "The Potential and Problems of Nuclear Power," lecture at the National Energy Forum, Washington, D.C., September 24, 1971. U.S. Atomic Energy Commission, Washington, 1971.
9. J. R. Schlesinger, "Power Production, Health and the Environment," lecture before the American Public Health Association, Atlantic City, New Jersey, November 14, 1972. U.S. Atomic Energy Commission, Washington, 1971.

CHAPTER 7

1. G. T. Seaborg and J. L. Bloom, "Fast Breeder Reactors," *Scientific American* **223**, No. 5, 13 (1970).
2. J. G. Yevick, Ed., *Fast Reactor Technology: Plant Design,* M.I.T. Press, Cambridge, Massachusetts, 1966.
3. Joint Committee on Atomic Energy Hearings, *Atomic Energy Commission Authorizing Legislation, Fiscal Year 1973,* Part 2, Government Printing Office, Washington, 1973.
4. Special Issue on Molten Salt Reactor Technology, *Nuclear Applications and Technology* **8**, No. 2 (1970).
5. M. W. Rosenthal et al., "Recent Progress in Molten Salt Reactor Development," *Atomic Energy Review* **9**, No. 3 (1971).
6. R. H. Simon and G. J. Schlueter, "From High Temperature Gas-Cooled Reactor to Gas-Cooled Fast Breeder Reactors," paper presented at the Joint Power Generation Conference, Boston, Massachusetts, September 10–14, 1972. Gulf-General Atomic Company, San Diego, California.
7. J. B. Dee, P. Fortescue, and J. A. Larrimore, "Gas-Cooled Fast Breeder Reactor Studies," paper presented at the Annual Meeting of the American Nuclear Society, Boston, Massachusetts, June 13–17, 1971. Gulf-General Atomic Company, San Diego, California.
8. R. Balent, "U.S. Liquid Metal Fast Breeder Reactor Demonstration Plant," paper presented at the Joint Power Generation Conference, Boston, Massachusetts, September 10–14, 1972. Atomics International, Canoga Park, California 91304.

CHAPTER 8

1. M. King Hubert, "The Energy Resources of Earth," *Scientific American* **224**, No. 3, 60 (1971).
2. C. Putnam, *Power from the Wind,* D. Van Nostrand Co., New York, N.Y., 1948.
3. W. E. Heronemus, "Power from Offshore Winds," Proceedings of the 8th Annual Marine Technology Society Conference, Washington, 1972.
4. W. E. Heronemus, "The United States Energy Crisis: Some Proposed Gentle Solutions," presented at the joint meeting of the American Society of Mechanical Engineers and The Institute of Electrical and Electronics Engineers, West Springfield, Massachusetts, January 12, 1972. Available from the author, University of Massachusetts, Amherst, Massachusetts 01002.
5. Clarence Zener, "Solar Sea Power," *Physics Today* **26**, No. 1, 48 (1973).
6. NSF/NASA Solar Energy Panel, "Solar Energy as a National Energy Resource," 1973. Available from the Department of Mechanical Engineering, University of Maryland, College Park, Maryland 20742.

CHAPTER 9

1. W. J. Hickel et al., *Geothermal Energy,* University of Alaska, College, Alaska 99701, 1972. A comprehensive assessment of geothermal energy.
2. United Nations, "United Nations Symposium on the Development and Utilization of Geothermal Resources," Pisa, Italy, Sept. 22-Oct. 1, 1970, Report ST/TAD/Series C/12b, New York, 1970. The standard reference.
3. L. T. Grose, "Geothermal Energy: Geology, Exploration, and Development," *Mineral Industries Bulletin* **14**, No. 6, 1 (1971); ibid. **15**, No. 1 (1972).
4. R. W. Rex, in *California Water,* D. Seckler, Ed. University of California Press, Berkeley, 1971.
5. M. Goldsmith, "Geothermal Resources in California, Potentials and Problems," Environmental Quality Laboratory Report No. 5, California Institute of Technology, Pasadena, California 91109, 1971.
6. D. N. Anderson and L. H. Axtell, Ed., "Geothermal Overviews of the Western United States," Geothermal Resources Council, Davis, California, 1972.
7. R. W. Rex, "Geothermal Energy, the Neglected Energy Option," *Bulletin of the Atomic Scientists* **27**, No. 8, 52 (1971).
8. Panel on Geothermal Energy Resources, "Assessment of Geothermal Energy Resources," U.S. Department of the Interior, Washington, 1972.
9. J. Barnea, "Geothermal Power," *Scientific American* **226**, No. 1, 70 (1972).
10. A. Kaufman, "The Economics of Geothermal Power in the United States," in "U.N. Symposium on the Development and Utilization of Geothermal Resources," Pisa, Italy, Sept. 22-Oct. 1, 1970, Report ST/TAD/Series C/12b, New York, 1970.
11. G. D. Bodvarsson, "Evaluation of Geothermal Prospects and the Objectives of Geothermal Exploration," *Geoexploration* **8**, 7 (1970).

CHAPTER 10

1. Proceedings of the United Nations Conference on New Sources of Energy, Rome, Italy, Aug. 21–31, 1961, United Nations, New York, 1964. The standard reference.

2. NSF/NASA Solar Energy Panel, "Solar Energy as a National Energy Resource," Department of Mechanical Engineering, University of Maryland, College Park, Maryland 20742, 1973.
3. H. Rau, *Solar Energy*, The Macmillan Co., New York, 1964.
4. R. Tybout and G. Lof, "Solar House Heating," *Natural Resources Journal* **10**, 268 (1970).
5. A. F. Hildebrandt, G. M. Haas, W. R. Jenkins, and J. P. Colace, "Large-Scale Concentration and Conversion of Solar Energy," *Eos* **53**, No. 7, 684 (1972).
6. E. A. Farber, W. H. Bussell, et al., "Solar Energy Studies," Technical Progress Report 9, Florida Engineering and Industrial Experiment Station, University of Florida, Gainesville, Florida 32601, 1960.
7. F. E. Edlin, "Worldwide Progress in Solar Energy," Proceedings of the 3rd Intersociety Energy Conversion Engineering Conference, New York, 1968.
8. H. Tabor, "Use of Solar Energy for Cooling Purposes," *Solar Energy* **6**, No. 4, 136 (1962).

9. A. M. Zarem and D. D. Erway, Eds., *Introduction to the Utilization of Solar Energy*, McGraw Hill, New York, 1963.
10. A. B. Meinel and M. P. Meinel, "Physics Looks at Solar Energy," *Physics Today* **25**, No. 2, 44 (1972).
11. F. Daniels and J. A. Duffie, *Solar Energy Research*, University of Wisconsin Press, Madison, 1955.
12. F. Daniels, "Energy storage problems," *Solar Energy* **6**, No. 3, 78 (1962).
13. F. Bennett, "Monthly Mass of Mean Daily Insolation for the United States," *Solar Energy* **9**, No. 3, 145 (1965).
14. G. O. G. Lof, J. A. Duffie, and C. O. Smith, "World distribution of solar radiation," *Solar Energy* **10**, No. 1, 27 (1966).
15. J. H. Anderson and J. H. Anderson, Jr., "Thermal Power from Sea-Water," *Mechanical Engineering* **88**, No. 4, 41 (1966).
16. C. J. Swet, "A Universal Solar Kitchen," Proceedings of the 7th Intersociety Energy Conversion Engineering Conference, San Diego, 1972. Applied Physics Laboratory, Johns Hopkins University, Silver Spring, Maryland.

CHAPTER 11

1. Ad Hoc Committee on Solar Cell Efficiency, *Solar Cells*, National Academy of Sciences, Washington, 1972.
2. F. Daniels, *Direct Use of the Sun's Energy*, Yale University Press, New Haven, 1964.
3. Special issue on Satellite Solar Power Station and Microwave Transmission to Earth, *Journal of Microwave Power* **5**, No. 4, 206 (1970).
4. P. E. Glaser, "Satellite Solar Power Station," *Solar Energy Journal* **12**, No. 3, 353 (1969).
5. B. Chalmers, H. E. LaBelle, Jr., and A. I. Mlavsky, "Growth of controlled profile crystals from the melt," *Materials Research Bulletin* **6**, 571 (1971).
6. P. E. Glaser, "Power from the Sun: Its Future," *Science* **162**, 857 (1968).
7. C. E. Backus, "A Solar-electric Residential Power System," Proceedings of the 7th Intersociety Energy Conversion Engineering Conference, San Diego, 1972. Arizona State University, Tempe, Arizona 85281.
8. E. L. Ralph, "A Plan to Utilize Solar Energy as an Electric-power Source," Proceedings of the Eighth Photovoltaic Specialists Conference, Seattle, Washington, 1970, Institute of Electrical and Electronics Engineers, New York. Catalogue No. 70C 32 ED.

9. A. Smith, "Status of Photovoltaic Power Technology," paper presented at the Winter Annual Meeting of the American Society of Mechanical Engineers, New York, 1968. Paper No. 68-WA/sol-1.
10. W. R. Cherry, "The Generation of Pollution-Free Electrical Power from Solar Energy," Transactions of the American Society of Mechanical Engineers, *Journal of Engineering for Power* **94**, Series A, No. 2, 78 (1972).
11. M. Wolf, "Historical Development of Solar Cells," 25th Power Sources Conference, Atlantic City, New Jersey, May 23–25, 1972. University of Pennsylvania, Philadelphia, Pennsylvania 19104.
12. C. G. Currin et al., "Feasibility of Low Cost Silicon Solar Cells," Proceedings of the 9th Photovoltaic Specialists Conference, Silver Spring, Maryland, May 2–4, 1972, Institute of Electrical and Electronics Engineers, New York.
13. K. W. Boer, "Large Scale Use of Photovoltaic Cells for Terrestrial Solar Energy Harvesting," Proceedings of the 9th Photovoltaic Specialists Conference, Silver Spring, Maryland, May 2–4, 1972, Institute of Electrical and Electronics Engineers, New York, 1972.

CHAPTER 12

1. L. L. Anderson, "Energy Potential from Organic Wastes: A Review of the Quantities and Sources," Bureau of Mines Information Circular 8549, Washington, 1972, $0.30.
2. E. G. Davis, L. L. Feld, and J. H. Brown,

"Combustion Disposal of Manure Wastes and Utilization of the Residue," Bureau of Mines Technical Progress Report 46, Washington, January 1972.

3. H. R. Appell, I. Wender, and R. D. Miller, "Conversion of Urban Refuse to Oil," Bureau of Mines Technical Progress Report 25, Washington, May 1970.

4. H. R. Appell, Y. C. Fu, S. Friedman, P. M. Yavorsky, and I. Wender, "Converting Organic Wastes to Oil: A Replenishable Energy Source," Bureau of Mines Report of Investigations 7560, Washington, 1971.

5. W. S. Sanner, C. Ortuglio, J. G. Walters, and D. E. Wolfson, "Conversion of Municipal and Industrial Refuse Into Useful Materials by Pyrolysis," Bureau of Mines Report of Investigations 7428, Washington, August 1970.

6. H. F. Feldman, "Pipeline Gas from Solid Wastes," *Chemical Engineering Applications in Solid Waste Treatment*, American Institute of Chemical Engineers Symposium Series **122**, 68 (1972).

7. H. F. Feldman, "Pipeline Gas From Solid Wastes," *Chemical Engineering Progress* **67**, 51 (1971).

8. C. B. Kenahan and E. P. Flint, "Bureau of Mines Research Programs on Recycling and Disposal of Mineral-, Metal-, and Energy-Based Solid Wastes," Bureau of Mines Information Circular 8529, Washington, 1971.

9. C. B. Kenahan, "Solid Waste—Resources Out of Place," *Environmental Science and Technology* **5**, 594 (1971).

10. D. E. Wolfson, J. A. Beckman, J. G. Walters, and D. J. Bennett, "Destructive Distillation of Scrap Tires," Bureau of Mines Report of Investigations 7302, Washington, September 1969.

11. G. M. Mallan and C. S. Finney, "New Techniques in the Pyrolysis of Solid Wastes," paper presented at 73rd National Meeting of the American Institute of Chemical Engineers, Minneapolis, Minnesota, August 1972. Garrett Research and Development Company, 1855 Carrion Road, La Verne, California 91750.

12. W. J. Oswald and C. G. Golueke, "Biological Transformation of Solar Energy," *Advances in Applied Microbiology*, Vol. II, Academic Press, New York, 1960.

13. NSF/NASA Solar Energy Panel, "An Assessment of Solar Energy as a National Energy Resource," College Park, Maryland, 1972. Contact Dr. Frederick H. Morse, Department of Mechanical Engineering, University of Maryland, College Park, Maryland 20742.

14. G. L. M. Christopher, "Biological Production of Methane from Organic Materials," United Aircraft Laboratories Research Report K910906-13, 1971. United Aircraft Corporation, East Hartford, Connecticut 06108.

CHAPTER 13

1. "Controlled Thermonuclear Research," Hearings before the Subcommittee on Research, Development, and Radiation of the Joint Committee on Atomic Energy, 92nd Congress, Nov. 10 and 11, 1971, Part II. Government Printing Office, No. 70-999 0, Washington, 1971, $1.50. A comprehensive reference.

2. David J. Rose, "Controlled Nuclear Fusion: Status and Outlook," *Science* **172**, 797 (1971).

3. Bruno Coppi and Jan Rem, "The Tokamak Approach in Fusion Research," *Scientific American* **227**, 65 (1972).

4. A. P. Fraas and H. Postma, "Preliminary Appraisal of the Hazards, Problems of a D-T Fusion Reactor Power Plant," Oak Ridge National Laboratory Report ORNL-TM-2822, Oak Ridge, Tennessee, 1970.

5. *Nuclear Fusion Reactors* [Proceedings of the British Nuclear Energy Society Conference on Fusion Reactors, United Kingdom Atomic Energy Authority (UKAEA), Culham Laboratory, September 17–19, 1969], J. L. Hall and J. H. C. Maple, Eds., Culham Laboratory, 1970.

6. D. J. Rose, "Engineering Feasibility of Controlled Fusion: A Review," *Nuclear Fusion* **9**, 183 (1969).

7. T. K. Fowler, "Fusion Research in Open-Ended Systems," *Nuclear Fusion* **9**, 3 (1969). A good review of magnetic mirror machines.

8. A. D. Beach et al., "Temperature and Density Measurements in the Mid-Plane of a Long Theta Pinch," *Nuclear Fusion* **9**, 215 (1969).

CHAPTER 14

1. Testimony of Robert L. Hirsch, "Controlled Thermonuclear Research," Hearings before the Subcommittee on Research, Development, and Radiation of the Joint Committee on Atomic Energy, 92nd Congress, Nov. 10 and 11, 1971, Part II. Government Printing Office No. 70-999 0, Washington, 1971, $1.50.

2. Moshe J. Lubin and Arthur P. Fraas, "Fusion by Laser," *Scientific American* **224**, 21 (1971).

3. L. A. Booth, "Central Station Power Generation by Laser-Driven Fusion," Los Alamos Scientific Laboratory Report LA-4858-MS, Vol. 1, Los Alamos, New Mexico, 1972.

4. A. P. Fraas, "The Blascon—An Exploding Pellet Fusion Reactor," Oak Ridge National Laboratory Report ORNL-TM-3231, Oak Ridge, Tennessee, 1971.

5. Lowell Wood and John Nuckolls, "Fusion Power," *Environment* **14**, No. 4, 29 (1972).

CHAPTER 15

1. H. Perry, "Conservation of Energy," Government Printing Office number 5270-0160, Washington, 1972, $0.50. A report prepared for the Senate Committee on Interior and Insular Affairs.

2. Electric Research Council, "The Electric Utilities Industry Research and Development Goals Through the Year 2000," New York, 1971.

3. R. B. Korsveyer, "Underground Air Storage and Electrical Energy Production,"

Oak Ridge National Laboratory Report number ORNL-NSF-EP-11, Oak Ridge, Tennessee, 1971.

4. H. C. Hottel and J. B. Howard, "An Agenda for Energy," *Technology Review* **74**, 38 (1972).

5. National Economic Research Associates, Inc., "Fuels for the Electric Utility Industry 1971–1985," Edison Electric Institute, New York, 1972.

CHAPTER 16

1. E. B. Forsyth, "Underground Power Transmission By Superconducting Cable," Brookhaven National Laboratory Report BNL 50325, Upton, New York 11973, 1972. A thorough and readable assessment of underground transmission technology, not limited to superconducting cables.

2. P. H. Rose, "Underground Power Transmission," *Science* **170**, 267 (1970).

3. James Nicol, "Underground Power Transmission," prepared for the Electric Research Council by Arthur D. Little, Inc., 1971. Available for $15 from Arthur D. Little, Inc., Cambridge, Massachusetts 02140. A 17 page summary of the study, "Planning for More Economical Underground Power Transmission," presented at the Institute of Electrical and Electronics Engineers Underground Transmission Conference, Pittsburgh, Pennsylvania, May 22–24, 1972 is also available from A. D. Little.

4. "Electric Utilities Industry Research and Development Goals Through the Year 2000," Electric Research Council Publication 1-71, 1971. Available for $8 from the Electric Research Council, 90 Park Avenue, New York, N.Y. 10016.

5. B. O. Pedersen, H. C. Doepken, Jr., and P. C. Bolin, "Development of a Compressed-Gas-Insulated Transmission Line," *IEEE Transactions on Power Applications and Systems* **90**, 2631 (1971).

6. R. W. Meyerhoff, "Superconducting Power Transmission," *Cryogenics* **11**, 91 (1971).

7. P. A. Klaudy, "Some Remarks on Cryogenic Cables," *Advances in Cryogenic Engineering* **11**, 684 (1966).

8. S. H. Minnich and G. R. Fox, "Cryogenic Power Transmission," *Cryogenics* **9**, 165 (1969).

9. S. H. Minnich and G. R. Fox, "Comparative Costs of Cryogenic Cables," presented at the Institute of Electrical and Electronics Engineers Winter Power Meeting, New York, January 1970. Paper Number 70, CP 169-PWR.

10. M. J. Jefferies, S. H. Minnich, and B. C. Belanger, "High Voltage Testing of a High-Capacity Liquid Nitrogen Cooled Cable," *IEEE Transactions on Power Applications and Systems* **92**, 514 (1973).

11. R. T. Allemann and B. M. Johnson, "An Evaporation-Cooled Underground Transmission System," presented to the Second Annual Thermal Power Conference, Washington State University, Pullman, October 6–8, 1971, page 499. Available from Battelle Pacific Northwest Laboratories, Richland, Washington.

CHAPTER 17

1. J. O'M. Bockris and S. Srinivasan, *Fuel Cells: Their Electrochemistry,* McGraw-Hill Book Company, St. Louis, Missouri, 1969.

2. H. A. Liebhafsky and E. J. Cairns, *Fuel Cells and Fuel Batteries, A Guide to Their Research and Development,* John Wiley & Sons, Inc., New York, 1968.

3. K. R. Williams, Ed., *An Introduction to Fuel Cells,* Elsevier Publishing Company, New York, 1966.

4. C. Berger, Ed., *Handbook of Fuel Cell Technology,* Prentice-Hall, Englewood, New Jersey, 1968.

5. M. F. Collins, R. Michalek, and W. Brink, "Design Parameters of a 300 Watt Ammonia-Air Fuel Cell System," Proceedings of the 7th IECEC, San Diego, 1972. Engelhard Minerals and Chemicals Corporation, Murray Hill, New Jersey 07974.

6. O. J. Adlhart, "The Phosphoric Acid Fuel Cell, A Long Life Power Source for the Low to Medium Wattage Range," Conference Proceedings of the 7th IECEC, San Diego, 1972. Engelhard Minerals and Chemicals Corporation, Murray Hill, New Jersey 07974.

7. Westinghouse Electric Corporation, "1970

Final Report Project Fuel Cell," Government Printing Office number 163.10:57, Washington, 1971, $6.75.

8. F. T. Bacon, "Fuel Cells, Past, Present, and Future." *Electrochimica Acta* **14**, 569 (1969).

CHAPTER 18

1. D. P. Gregory, D. Y. C. Ng, and G. M. Long, "The Hydrogen Economy," in *Electrochemistry of Cleaner Environments*, J. O'M. Bockris, Ed., Plenum Publishing Corporation, New York, 1972.

2. Harold Sorenson, "The Boston Reformed Fuel Car—A Low Polluting Gasoline Fuel System for Internal Combustion Engines," Proceedings of the 7th IECEC, San Diego, 1972. International Materials Corporation, Burlington, Massachusetts.

3. R. L. Costa and P. G. Grimes, "Electrolysis as a Source of Hydrogen and Oxygen," *Chemical Engineering Progress* **63**, 56 (1967).

4. M. R. Swain and R. R. Adt, Jr., "The Hydrogen-Air Fueled Automobile," Proceedings of the 7th IECEC, San Diego, 1972. School of Engineering and Environmental Design, University of Miami, Coral Gables, Florida 33124.

5. G. DeBeni and C. Marchetti, "Hydrogen, Key to the Energy Market," *Euro Spectra*, **IX**, No. 2, June 1970.

6. R. G. Murray, R. J. Schoeppel, and C. L. Gray, "The Hydrogen Engine in Perspective," Proceedings of the 7th IECEC, San Diego, 1972. R.J.S., Oklahoma State University, Stillwater, Oklahoma 74074.

7. Lawrence W. Jones, "Liquid Hydrogen as a Fuel for the Future," *Science* **174**, 367 (1971).

8. D. P. Gregory, "A New Concept in Energy Transmission," *Public Utility Fortnightly* **89**, 21 (1972).

9. K. V. Kordesch, "Hydrogen-Air/Lead Battery Hybrid System for Vehicle Propulsion," *Journal of the Electrochemical Society* **118**, 812 (1971).

10. W. E. Winsche, K. C. Hoffman, and F. J. Salzano, "Economics of Hydrogen Fuel for Transportation and Other Residential Applications," Proceedings of the 7th IECEC, San Diego, 1972. Brookhaven National Laboratory, Upton, New York 11973.

11. John V. Becker, "Prospects for Actively Cooled Hypersonic Transports," *Astronautics and Aeronautics* **9**, 32 (1971).

12. R. D. Witcofski, "Potentials and Problems of Hydrogen Fueled Supersonic and Hypersonic Aircraft," Proceedings of the 7th IECEC, San Diego, 1972. NASA Langley Research Center, Hampton, Virginia 33365.

13. J. J. Reilly and R. H. Wiswall, Jr., "Metal Hydrides for Energy Storage," Proceedings of the 7th IECEC, San Diego, 1972. Brookhaven National Laboratory, Upton, New York 11973.

14. J. J. Reilly and R. H. Wiswall, Jr., *Inorganic Chemistry* **6**, 2220 (1967); **7**, 2254 (1968); and **9**, 1678 (1970).

15. E. C. Ashby, G. J. Brendel, and H. E. Redman, *Inorganic Chemistry* **2**, 499 (1963).

16. D. P. Gregory, D. Y. C. Ng, and G. M. Long, "Electrolytic Hydrogen as a Fuel," Institute of Gas Technology, Chicago, Illinois 60616, January 1971.

17. F. A. Martin, "The Safe Distribution and Handling of Hydrogen for Commercial Application," Proceedings of the 7th IECEC, San Diego, 1972. Linde Division, Union Carbide Corporation, New York, New York 10017.

18. D. P. Gregory and J. Wurm, "Production and Distribution of Hydrogen as a Universal Fuel," Proceedings of the 7th IECEC, San Diego, 1972. Institute of Gas Technology, Chicago, Illinois 60616.

19. J. E. Mrochek, "Economics of Hydrogen and Oxygen Production by Water Electrolysis and Competitive Processes," *Abundant Nuclear Energy*, W. W. Gregorieff, Ed., U.S. Atomic Energy Commission, Washington, 1969.

20. B. Eastlund and W. C. Gough, "Generation of Hydrogen by UV Radiation," paper presented at Symposium on Non-Fossil Chemical Fuels, 163rd National Meeting of the American Chemical Society, Boston, April 1972.

21. E. C. Tanner and R. A. Huse, "A Hydrogen-Electric Utility System with Particular Reference to Fusion as the Energy Source," Proceedings of the 7th IECEC, San Diego, 1972. E.C.T., Plasma Physics Laboratory, Princeton University, Princeton, New Jersey 08540.

22. W. Hausz, G. Leeth, and C. Meyer, "Eco-Energy," Proceedings of the 7th IECEC, San Diego, 1972. Tempo, General Electric Company, Center for Advanced Studies, Santa Barbara, California 93101.

23. J. R. Bartlit, F. J. Edeskuty, and K. D. Williamson, Jr., "Experience in Handling, Transport and Storage of Liquid Hydrogen—The Recyclable Fuel," Proceedings of the 7th IECEC, San Diego, 1972. Los Alamos Scientific Laboratory, Los Alamos, New Mexico 87544.

24. J. O'M. Bockris, "A Hydrogen Economy," *Science* **176**, 1323 (1972).

25. Derek P. Gregory, "The Hydrogen Economy," *Scientific American* **228**, 13 (1973).

26. C. Marchetti, "Hydrogen and Energy," *Chemical Economy & Engineering Review* **5**, 7 (1973).

CHAPTER 20

1. Stanford Research Institute, "Patterns of Energy Consumption in the United States," Office of Science and Technology, Washington, 1972.
2. J. C. Moyers, "The Value of Thermal Insulation in Residential Construction," Oak Ridge National Laboratory, Report ORNL-NSF-EP-9, Oak Ridge, Tennessee, 1971.
3. R. G. Stein, "Architecture and Energy," lecture delivered at the American Association for the Advancement of Science, Philadelphia Meeting, December 29, 1971. R. G. Stein and Associates, 588 Fifth Ave., New York, N.Y.
4. Earl Cook, "The flow of energy in an industrial society," *Scientific American* **224**, No. 3, 134 (1971).
5. Office of Emergency Preparedness, "The Potential for Energy Conservation," Government Printing Office number 4102-00009, Washington, 1972, $2.00.
6. B. Hanon, "Bottles, Cans, Energy," *Environment* **14**, No. 2, 11 (1972).
7. C. Berg, "Energy Conservation Through Effective Utilization," National Bureau of Standards, 1972, unpublished.
8. Harry Perry, "Conservation of Energy," Government Printing Office number 5270-01602, Washington, 1972, $0.50. Prepared for the Senate Committee on Interior and Insular Affairs.
9. E. Hirst, "Energy Consumption for Transportation in the U.S.," Oak Ridge National Laboratory, Report ORNL-NSF-EP-15, Oak Ridge, Tennessee, 1972.
10. D. G. Harvey and J. A. Kudrick, "Minimization of Residential Energy Consumption," Proceedings of the 7th Intersociety Energy Conversion Engineering Conference, San Diego, 1972.
11. E. Hirst and J. Moyers, "Efficiency of Energy Use in the United States," *Science* **179**, 1299 (1973).
12. D. P. Gregory, "A Techno-Economic Study of the Cost-Effectiveness of Methods of Conserving the Use of Energy," Institute of Gas Technology, Chicago, Illinois 60616, June 1971.
13. A. B. Makhijani and A. J. Lichtenberg, "Energy and Well-Being," *Environment* **14**, No. 5, 10 (1972).

CHAPTER 21

1. D. Chapman, T. Tyrell, T. Mount, "Electricity Demand Growth and the Energy Crisis," *Science* **178**, 703 (1972).
2. W. E. Mooz and C. C. Mow, "California's Electric Quandary: Estimating Future Demand," Rand Corporation, Santa Monica, California 90406, 1972.
3. R. D. Doctor et al., "California's Electric Quandary: Slowing the Growth Rate," Rand Corporation, Santa Monica, California 90406, 1972.
4. W. Heller, "Coming to Terms with Growth and the Environment," in *Energy, Economic Growth and the Environment,* S. H. Schurr, Ed., Johns Hopkins Press, Baltimore, 1972.
5. P. Balestra, *The Demand for Natural Gas in the United States,* North-Holland, Amsterdam, 1967.
6. U.S. Federal Power Commission, *The 1970 National Power Survey,* Government Printing Office, Washington, 1971.
7. Committee on U.S. Energy Outlook, *U.S. Energy Outlook,* National Petroleum Council, Washington, 1972.

INDEX

Shell International Petroleum Company, 19–21, 123
Sheridan, E. T., 167
Silicon (*see* Photovoltaic cells)
Simon, R. H., 170
Sludge, 78, 157, 164
Smith, A., 171
Smith, C. O., 171
Smith, Morton C., 58, 59
Sodium, Liquid, 34, 36, 41, 42
Solar collectors, 62, 64, 65
Solar energy (*see also* Photovoltaic cells), 47–53, 61–71, 97, 117, 121, 134, 143, 147, 148, 150, 151
  Commercial prospects, 61, 62, 66
  Cost, 52, 61, 62, 66
  Efficiency, 63
  Environmental problems, 66
  Funding, 61
  Generation of electricity, 63–66
  Potential, 47, 48, 61
  Principle of operation, 62, 64–66
  Problems, 62
Solvent refining, 9, 165
Sorenson, Harold, 119, 173
Southern Methodist University, 58
Space conditioning, 61–63, 66, 113, 122, 131–134
Sporn, P., 169
Squires, Arthur M., 20–23, 168
Srinivasan, S., 173
Stanford Research Institute, 12, 174
Stanford University, 27, 106
Starr, Chauncey, 48
Steam turbines (*see* Turbines)
Stein, Richard G., 134, 135, 174
Steiner, D., 167
Storage of energy (*see* Energy, Storage)
Strip mining (*see* Mining)
Subbituminous (*see* Coal)
Sulfur (*see also* Pollution, Air), 11, 12, 14, 20, 60
  In coal, 5, 7, 15, 17, 27, 162
  In oil, 3, 19, 74–76
Sulfur hexafluoride, 104
Sulfur oxides (*see* Pollution, Air)
Superconducting magnets, 26, 28, 80, 81
Superconducting power lines (*see* Electric power transmission)
Suttle, Robert, 113
Swain, M. R., 174
Swet, C. J., 171
Sykes, J. A., Jr., 169
Synthane process (*see* Gasification of coal)
Synthesis gas, 12, 14, 19, 165
Synthetic fuels (*see also* Hydrogen *and* Synthetic natural gas), 48, 73–78, 124, 148
  Alcohol, 78
  Bioconversion (*see also* Anaerobic digestion), 73–75, 77
  BuMines process for oils from waste, 74–76
  Commercial prospects, 74, 76, 78
  Cost, 76, 77
  Efficiency, 74, 75
  Funding, 74, 76–78
  Garrett process for oil from wastes, 75–77

Hydrogenation of organic wastes, 73–76
  Potential, 73
  Pyrolysis (*see also* Pyrolysis), 73, 75–77
Synthetic natural gas (*see also* Gasification), 11, 14, 17, 124, 150, 157, 161, 163, 165
  Cost, 12

Tabor, Harry, 64, 171
Tanner, E. C., 174
TARGET (Team to Advance Research for Gas Energy Transformation), 112, 113
Tennessee, University of, Space Institute, 27
Tennessee Valley Authority, 34
Texaco Development Corporation, 19
Texas A&M University, 133
Texas Eastern Transmission Corporation, 118
Thermal breeder reactors (*see* Breeder reactors)
Thermal pollution (*see* Pollution, Thermal)
Thorium, 31, 34, 39, 158, 165
Tidal power, 46, 50
  Potential, 47, 48
TNT, 154, 155
Tokamaks (*see* Fusion, Magnetic containment)
Tomlinson, L. O., 168
Torrax Systems Inc., 77
Transmission of energy (*see* Energy, Transmission, Electric power transmission, *and* Hydrogen)
Tritium, 52, 79–82, 91, 94, 160, 161, 165
  Hazards, 82–84
Turbines, 52, 63, 66, 158, 165
  Gas, 17–19, 21, 23, 28, 35, 137, 158, 162
    Efficiency, 18, 111, 112, 136
  Isobutane, 19, 57, 59
  Steam, 18, 19, 23, 26, 34, 56, 57, 64, 90, 91, 102, 114, 137, 158, 162
    Efficiency, 27, 110–112, 136
Tybout, R., 170
Tyco Laboratories, 68
Tyrrell, T., 140, 142, 175

Union Carbide Corporation
  Linde division, 106
  Parma Research Center, 118
  Tarrytown Research Center, 77
Union of Concerned Scientists, 35
Union Oil Company, 56
United Aircraft Corporation
  Corporate Research Laboratories, 18, 21, 77, 78, 168
  Pratt & Whitney Aircraft division, 109–113, 115
United Nations, 64, 170
United States Government
  Atomic Energy Commission, 31–33, 35–39, 42, 44, 58, 82, 84, 87, 91, 101, 107, 121, 122, 150, 169
    Controlled Thermonuclear Research division, 80, 81
    Military Applications division, 88
  Department of Commerce
    Bureau of Economic Analysis, 141
    National Bureau of Standards, 133
  Department of Housing and Urban Development, 134

The American Association for the Advancement of Science is the major general-scientific organization in the U.S. It was organized in 1848 as a private, non-profit association.

The AAAS has more than 125,000 individual members, most of whom are professional scientists working in universities, corporations and government agencies. The membership also includes many amateur scientists and persons who are simply interested in science.

In addition, the AAAS is the world's largest federation of scientific organizations. More than 290 groups, representing the entire spectrum of scientific disciplines, are affiliated with the Association.

The objectives of AAAS are:
- To further the work of scientists and to facilitate cooperation among them.
- To improve the effectiveness of science in the promotion of human welfare.
- To increase public understanding and appreciation of the importance and promise of the methods of science in human progress.

ACKNOWLEDGMENTS

The authors would like to thank Susan K. Carhart for copy editing, William Clipson and Pamela Stevenson for preparing the drawings, Joyce Richards for proofreading, and Fanny Groom for typing. Special thanks are also due Robert Potter for his initial encouragement and continuing enthusiasm and the staff of *Science* for their help in preparing the original articles.

Designed by Gerard A. Valerio, EditaGraphics, Inc.

Composed in Linotype Times Roman and Franklin Gothic by the
Monotype Composition Company

Printed letterpress on 55 pound Dontext and bound in Holliston Zeppelin bookcloth
by the R. R. Donnelly Printing Company